The Garbage Crisis

A Global Challenge for Engineers

Synthesis Lectures on Engineers, Technology, and Society

Editor
Caroline Baillie, *University of Western Australia*

The mission of this lecture series is to foster an understanding for engineers and scientists on the inclusive nature of their profession. The creation and proliferation of technologies needs to be inclusive as it has effects on all of humankind, regardless of national boundaries, socio-economic status, gender, race and ethnicity, or creed. The lectures will combine expertise in sociology, political economics, philosophy of science, history, engineering, engineering education, participatory research, development studies, sustainability, psychotherapy, policy studies, and epistemology. The lectures will be relevant to all engineers practicing in all parts of the world. Although written for practicing engineers and human resource trainers, it is expected that engineering, science and social science faculty in universities will find these publications an invaluable resource for students in the classroom and for further research. The goal of the series is to provide a platform for the publication of important and sometimes controversial lectures which will encourage discussion, reflection and further understanding.

The series editor will invite authors and encourage experts to recommend authors to write on a wide array of topics, focusing on the cause and effect relationships between engineers and technology, technologies and society and of society on technology and engineers. Topics will include, but are not limited to the following general areas; History of Engineering, Politics and the Engineer, Economics , Social Issues and Ethics, Women in Engineering, Creativity and Innovation, Knowledge Networks, Styles of Organization, Environmental Issues, Appropriate Technology

The Garbage Crisis: A Global Challenge for Engineers
Randika Jayasinghe, Usman Mushtaq, Toni Alyce Smythe, and Caroline Baillie
2013

Engineers, Society, and Sustainability
Sarah Bell
2011

Engineering and Social Justice
Donna Riley
2008

Engineering, Poverty, and the Earth
George D. Catalano
2007

Engineers within a Local and Global Society
Caroline Baillie
2006

Globalization, Engineering, and Creativity
John Reader
2006

Engineering Ethics: Peace, Justice, and the Earth
George D. Catalano
2006

The Garbage Crisis: A Global Challenge for Engineers
Randika Jayasinghe, Usman Mushtaq, Toni Alyce Smythe, and Caroline Baillie

ISBN: 978-3-031-00983-9   paperback
ISBN: 978-3-031-02111-4   ebook

DOI 10.1007/978-3-031-02111-4

A Publication in the Springer Nature series
SYNTHESIS LECTURES ON ADVANCES IN AUTOMOTIVE TECHNOLOGY

Lecture #18
Series Editor: Caroline Baillie, *University of Western Australia*
Series ISSN
Synthesis Lectures on Engineers, Technology, and Society
Print 1933-3633   Electronic 1933-3641

The Garbage Crisis

A Global Challenge for Engineers

Randika Jayasinghe
University of Western Australia

Usman Mushtaq
Queen's University, Canada

Toni Alyce Smythe
Department of Water, Perth

Caroline Baillie
University of Western Australia

SYNTHESIS LECTURES ON ENGINEERS, TECHNOLOGY, AND SOCIETY #18

ABSTRACT

This book will focus on *"Waste Management,"* a serious global issue and the engineers' responsibility towards finding better solutions for its sustainable management. Solid waste management is one of the major environmental burdens in both developed and developing countries alike. An alarming rate of solid waste generation trends can be seen as a result of globalization, industrialization, and rapid economic development. However, low-income and marginalized sectors in society suffer most from the unfavorable conditions deriving from poor waste management. Solid waste management is not a mere technical challenge. The environmental impact, socio-economic, cultural, institutional, legal, and political aspects are fundamental in planning, designing, and maintaining a sustainable waste management system in any country. Engineers have a major role to play in designing proper systems that integrate stakeholders, waste system elements, and sustainability aspects of waste management. This book is part of a focused collection from a project on Engineering and Education for Social and Environmental Justice. It takes an explicitly social and environmental justice stance on waste and attempts to assess the social impact of waste management on those who are also the most economically vulnerable and least powerful in the society. We hope that this book will assist our readers to think critically and understand the framework of socially and environmentally just waste management.

KEYWORDS

waste management, social and environmental justice, social impact, marginalized sectors, engineering and social justice

Contents

4 Waste Management in the Global North 69

Randika Jayasinghe

5 Waste Management in the Global South: A Sri Lankan Case Study 87

Randika Jayasinghe

6 Assessing the Feasibility of Waste for Life in the Western Province of Sri Lanka 109

Toni Alyce Smythe

Acknowledgments

This book is an outcome of an Australian research project (Engineering Education for Social and Environmental Justice) funded by the Federal Government's Office of Teaching and Learning (CG10-1519).

We would like to extend our special thanks to Adam Johnson, CEO of the Western Municipal Regional Council, for his valuable time on proofreading and editing the manuscript and for his constructive comments and suggestions.

We would also like to thank Dr. Rita Armstrong, for her guidance and valuable support in completing this book.

Last but not least, we thank everyone who shared their ideas and knowledge with us in many ways. Their contributions have served as valuable sources of information for this book.

Randika Jayasinghe, Usman Mushtaq, Toni Alyce Smythe, and Caroline Baillie
January 2013

Figure Credits

Figure 3.2	used courtesy of Stop Dump Site 41. `http://stopdumpsite41.ca`
Figures 3.3, 3.4	photographs courtesy of Council of Canadians. `http://www.canadians.org/`
Figures 6.1, 6.2	photographs courtesy of Eric Feinblatt.

CHAPTER 1

Introduction

Randika Jayasinghe

"A good solid waste system is like good health: if you are lucky to have it, you don't notice it, it is just how things are, and you take it for granted. On the other hand, if things go wrong, it is a big and urgent problem and everything else seems less important."(pg. 1)

UN-HABITAT [1]

1.1   INITIAL THOUGHTS

In the early years of human settlement, the small number of people in a community and the scarcity of items meant that there was very little waste. Natural resources were not exploited extensively due to limited technology available at the time. People re-used and repaired their belongings, while organic wastes were fed to livestock and used as fertilizer for crops. However, populations have grown, so have the problems associated with waste. The earth will soon be saturated by human population. The United States Census Bureau (USCB) has estimated that the world population exceeded 7 billion in 2012 [2]. According to current projections, there will be a continued increase in population with the global population expected to reach between 7.5 and 9.5 billion by 2050 [3]. Hence, a better understanding of the scarcity of natural resources compared to the growing quantities of waste produced by human activities is required.

Waste management is a worldwide problem in poor and rich cities alike. Managing waste in a socially and environmentally acceptable manner is one of the key challenges of the 21st century. We are well aware of the growing waste problem that is literally choking the world we live in today. While the growing waste problem is a popular subject in the media, which regularly reports on "garbage crises," "environmental pollution," and "landfills," many people do not think about waste after disposing of it from their homes. Thus, individuals have very little awareness of where their waste ends up.

When considering waste management in our communities, it is important to ask the following questions.

- Who handles waste?

- Where are the final destinations of this waste? Is it an engineered landfill or an open dumpsite?

- Will waste be separated to recover any recyclable materials? If so, who will do that?

- Where are hazardous and electronic wastes dumped?

People tend to put the blame on the authorities for the inefficiencies of the services, lack of both material and financial resources, and relate the issue to rapid population growth and urbanization [4]. These are issues of importance. However, if only these issues are addressed, the key elements that need urgent attention may be missed; issues that are directly related to the social aspects of waste management.

1.2   WASTE MANAGEMENT IN THE WORLD

Waste management in today's context represents a system with a standard technical idea; removal of waste from one area and disposal in another place [5]. This process creates appropriate intermediate or final destinations for different waste components according to their sources, value, and physical state. The remainder, which still makes up a considerable portion of the total waste, ends up in landfills, open dumps, buried or thrown into waterways. We think of finding a solution to our waste when these improper disposal practices create problems such as polluted air, contaminated water sources and over flowing landfills. According to Clapp (2002), the main issue with waste management is the growing distance between waste and its producers. When consumers have little understanding of the ecological and social impacts of waste, there is little motivation to change their wasteful habits [4].

Bearing this in mind, we present waste management as an iceberg model. In this model, the waste problem we see is represented as the tip of the iceberg. What is underneath the water surface, the unseen, is the actual problem, so it requires greater focus. This unseen layer of the iceberg represents the social and environmental determinants of waste management. We need to continually ask "why" and "what needs to be done" for both layers of the iceberg and look beyond the tip of the iceberg that we generally perceive.

This book presents a series of waste management challenges faced in everyday life. We have looked at each of these situations from different angles, but using the same critical lens—a lens of social and environmental justice.

1.2.1 WASTE AS A SOCIAL DILEMMA

When discussing waste as a problem, we usually emphazise the health issues, cleanliness, aesthetics, or environmental damage such as pollution and contamination. An essential feature of the way we look at the waste problem is that we tend to neglect other social aspects linked with waste management, which are often overlooked by modern society and considered as something that does not need urgent attention. Negative connotations related to waste often impede open discussion in this area and leads to those in the lower levels of the waste economy, working with waste for their survival, having low social status and little influence over the decision-making process.

Waste is not only a worldwide environmental problem; it is also a worldwide social problem. According to Zurbrügg (2003), the poor in society suffer most from the unfavorable conditions derived from disorganized waste management [6]. Local authorities allocate more resources to manage waste in affluent areas where citizens with more financial and political power live. They often ignore or pay less attention to waste management in areas where the majority of the poor people live. A lack of environmentally acceptable solid waste management has created significant public health risks, particularly for the individuals who live in close proximity to waste dumps and make their living through scavenging activities to supplement their unsatisfactory earnings [7].

Franklin (1992) in her book, *The Real World of Technology,* suggests, when deciding upon a particular project, not to simply consider benefits and costs but to question "who benefits and who pays?" [8]. According to her, it is important to consider and question the following aspects before implementing a project [9]. Does the project:

- promote justice;

- restore mutual benefit;

- confer divisible or indivisible benefits (according to Franklin, working on ways to improve the environment is an indivisible benefit—all groups of people benefit);

- favor people over machines;

- promote conservation over waste; and

• favor reversible over irreversible.

She further suggests that we should do cost accounting for social, environmental, and economic aspects, and that we must continually ask the same question, "Who benefits and who pays?," for each of these areas. Her thinking could serve us well in our discussion on waste management.

1.2.2   WASTE—SHIFTING THE RESPONSIBILITY

Waste often follows the path of least resistance and ends up in marginalized communities throughout the world, in countries both rich and poor alike. According to Pearce and Warford (1993), "it is perfectly possible for a single nation to secure sustainable development, in the sense of not depleting its own stock of capital assets, at the cost of procuring unsustainable development in another country" [10, p. 185]. This statement clearly applies to waste management in today's context. As Clapp (2002) points out, movement of waste often takes advantage of economic inequalities making their ways to disadvantaged communities [4]. As powerful communities externalize their environmental impacts, waste ends up in remote areas and in the least powerful neighborhoods [11]. According to Bullard (2001), these communities are given a false choice of "no jobs no development" vs. "risky low paying jobs and pollution" [12, p. 167]. In this book, we explore this idea of shifting the responsibility of managing waste to people in other places. As Rothman (1998) stated, "We must understand that solving environmental problems means more than handing them off, i.e., passing the buck to people in other places or in other times. It must be understood what we mean by growth, i.e., that we make a clear distinction between economic growth and growth in human well-being" [10, p. 191].

1.2.3   WASTE—A RESOURCE OUT OF PLACE

Waste can be considered a "resource out of place." If managed properly, it can provide material or energy in useful forms. A better system would incorporate reduction of waste to be disposed at the source, recovery of energy or material and safe disposal of what is left causing minimum risk to the environment and to the society. The recovery of waste can contribute to alleviating problems such as poverty, unemployment, and inadequate waste disposal. If economically marginalized populations of the world are to benefit, the range of "small-scale, low-cost and environmentally sound options" needs to be developed and implemented [13, p. 10]. Waste management is not a mere technical challenge. Understanding the environmental, socio-cultural, economic, legal, and political aspects are fundamental in planning, designing, and maintaining a sustainable waste management system in any country. It is hoped that this book will contribute to these efforts.

1.2.4   WASTE—THROUGH A "POST-DEVELOPMENT" FRAMEWORK

According to the post-development scholar Ferguson (1990), "development" as seen by developed nations, is not the only available form of engagement with problems of poverty, hunger, and oppression in developing nations [14]. He does not reject the idea of developers from the North working together with people from the South; however, he points out that developers should understand the politics surrounding the particular issues in the local context they are operating [15]. This takes into consideration the idea that it is not the "North" patronizing the "South," but a more collaborative approach is required.

Waste for Life (WfL) is a not-for-profit organization which follows the same approach; developing more collaborative ventures and sharing knowledge with the marginalized communities working with waste [16]. WfL is a loosely joined network of professionals, cooperatives, artists, and students of different disciplines and countries who work with waste collecting cooperatives to co-create, apply, and disseminate poverty-reducing technologies for repurposing waste [9]. A case study carried out to assess the needs and feasibility of implementing a WfL project is presented as the last chapter of this book.

1.2.5   ENVIRONMENTALLY AND SOCIALLY JUST WASTE MANAGEMENT

Waste management has focused historically on public health and environment. In many countries, this remains unchanged. We have taken a different path to this traditional approach, making an effort to include social aspects into the waste management system thereby focusing on the social values in the foreground, with environmental benefits as background. Bookchin (1993), a major figure in Social Ecology, has stated that most of the current ecological problems arise from societal and social problems and argues that ecological problems cannot be understood or solved without understanding and dealing with the problems within the society [17]. According to Bookchin, "to separate ecological problems from social problems......would be to grossly misconstrue the sources of the growing environmental crisis" [17, p. 1]. Hence, in this book, we have used a lens of social and environmental justice to analyze the waste problem. Throughout the book, we will be focusing on waste through this lens, more or less explicitly.

This book is not intended to provide answers to questions on technical and economic aspects of waste management. Instead, the book is intended to assist readers to think critically and understand the framework of socially and environmentally just waste management. Developing an environmentally and socially just system encourages us to analyze the complex interactions of all the stakeholders, including the most vulnerable parties at the bottom of the waste economy.

1.3   DEFINITION OF DEVELOPMENTAL STATUS

The Global North refers to the 57 countries with high human development that have a Human Development Index above 0.8 as reported in the United Nations Development Programme Report 2005. Most, but not all, of these countries are located in the Northern Hemisphere.

The Global South refers to the countries of the rest of the world, most of which are located in the Southern Hemisphere. It includes both countries with medium human development (88 countries with an HDI less than 0.8 and greater than 0.5) and low human development (32 countries with an HDI of less than 0.5). Most of the Global South is located in South and Central America, Africa, and Asia.

In this book, the terms "developed" and "Global North" are used synonymously. Similarly, the terms "developing" and "Global South" are used synonymously.

1.4 STRUCTURE OF THE BOOK

This book provides a critical analysis of the importance in managing waste in a socially and environmentally just manner. It develops a lens for viewing the typical waste management systems in different regions of the world, while at the same time encouraging readers to think how the impacts of waste are uneven and how this can be addressed.

The next chapter introduces the social and environmental justice framework. It is explicitly used as a framework for Chapter 3, which explores a landfill project in Canada, but also provides a general overview to explain the lens used to study waste throughout the book. Chapter 4 discusses waste management in the global North taking specific waste management scenarios as examples. In Chapter 5, the waste management in the global South is discussed taking Sri Lanka as a case study. Finally, Chapter 6 presents a case study carried out to assess the feasibility of implementing a WfL project in Sri Lanka.

1.5 FINAL THOUGHTS

Waste management is one of the major issues of engineering for the decades to come. There are no easy answers for any of the socio-environmental issues of waste management and there are no perfect solutions. We need to be critical and creative to find and to adapt viable solutions that will work in a particular situation. This can make a difference in the way the costs and benefits of waste management are currently distributed. We hope that this book will inspire readers to think critically, using a different lens to look at, analyze, and reflect upon waste management in cities worldwide.

REFERENCES

[1] UN-HABITAT, *Solid Waste Management in the World's Cities: Pre-publication presentation*, D.C. Wilson, A. Scheinberg, and L. Rodic-Wiersma, Editors. 2009, UNON Print Shop: Nairobi. 1

[2] U.S. Census Bureau, *World POPClock Projection*. International Data Base 2012. 1

[3] United States Census Bureau, *International Progrmas: World Population*, 2012, U.S. Dept of Commerce: US. 1

[4] Clapp, J., *The distancing of waste: Overconsumption in a global economy,*, in *Confronting Consumption*, T. Princen, M. Maniates, and K. Conca, Editors. 2002, MIT Press: Cambridge, p. 155–176. 2, 4

[5] Scheinberg, A., *A Bird in the Hand: Solid Waste Modernisation, Recycling and the Informal Sector*, in *Solid Waste Planning in the Real World, CWG-Green Partners Workshop* 2008: Cluj, Romania. 2

[6] Zurbrügg, C., *Urban Solid Waste Management in Low-Income Countries of Asia - How to Cope with the Garbage Crisis.* in *Scientific Committee on Problems of the Environment (SCOPE), Urban Solid Waste Management Review Session.* 2003. Durban, South Africa. 3

[7] Environmental Foundation LTD., *Climbing out of the Grabage Dump - Managing colombo's solid waste problem.* 2007; Available from: `http://www.efl.lk/publication/`. 3

[8] Franklin, U.M., *The real world of technology.* CBC Massey Lectures1992, Concord, Ontario: Anansi Publishers. 3

[9] Baillie, C., et al., *Needs and Feasibility: A Guide for Engineers in Community Projects——The Case of Waste for Life.* Synthesis Lectures on Engineers, Technology and Society 2010: Morgan & Claypool Publishers. 3, 5

[10] Rothman, D., S., *Environmental Kuznets curves——real progress or passing the buck? A case for consumption-based approaches.* Ecological Economics, 1998. **25**: p. 177–194. 4

[11] Ackerman, F. and S. Mirza, *Waste in the inner city: Asset or assault?* Local Environment, 2001. **6**(2): p. 113–120. 4

[12] Bullard, R.D., *Environmental Justice in the 21st Century: Race Still Matters.* Phylon (1960-), 2001. **49**(3/4): p. 151–171. 4

[13] WASTE Consultants, *Plastic Waste: Options for small-scale resource recovery (Urban Solid Waste Series 2)*, ed. I. Lardinois and A. van der Klundert1995, Amsterdam: TOOL Publications. 4

[14] Ferguson, J., *The anti-politics machine: "development," depoliticization and bureaucratic power in Lesotho*1990, Cambridge: Cambridge University Press. 4

[15] Ferguson, J. and L. Lohmann, *The Anti-politics machine: "Development" and Bureaucratic power in Lesotho.* The Ecologist, 1994. **24**(5): p. 176–181. 5

[16] Baillie, C., *Waste for Life: Socially just materials research*, in *Engineering and Social Justice : In the university and beyond*, C. Baillie, A.L. Pawley, and D. Riley, Editors. 2012, Purdue University press. p. 87–106. 5

[17] Bookchin, M., *What Is Social Ecology?* 1993; Available from: `http://www.anarchija.lt/`. 5

CHAPTER 2

Towards a Just Politics of Waste Management

Usman Mushtaq

THYME: I was right the first time, Christ! this ain't no love present. This is payoff, bribe, this is our kids running, sucking in toxic air cuz the white neighborhoods sure didn't wanna be looking at it, this plant takes care of the entire West Side, a more central locale would've made better sense, fifty blocks south of here. But that place is six-figured and white uh uh! And if we complain, stench, noise they'll say "minor inconvenience," "necessary tradeoff." (p. 56)

Kia Corthron [18]

2.1 INTRODUCTION

In Corthron's play [18], the main character, Thyme, is three months pregnant in environmentally poisoned Harlem while her husband, Erry, has a job that is slowly damaging his health. Through these characters, the play explores the race politics of pollution: where does pollution get placed, who is most affected by it, and how are those decisions made. The picture drawn through the characters is pessimistic. Black communities in the inner-city bear the brunt of pollution generated by white upper-class neighborhoods. They have little to no say in environmental decisions that not only affect them but also their children. Children in those neighborhoods have higher risks to their health and well-being. In the end, Erry passes away because of his exposure to lead in his workplace leaving Thyme to raise their child in an increasingly toxic environment.

This story of environmental racism is repeated in the siting and management of waste in North America. Communities of color are more likely to be on the "losing" side of environmental decisions than white communities [12]. Dumps are more often placed in neighborhoods with predominantly people of color. In 1987, three of the five largest commercial waste facilities were sited in predominantly black or hispanic neighborhoods while three out every five black/hispanic people lived in communities with unregulated/abandoned waste sites [23]. A follow-up study in 2007 found that "racial disparities in the distribution of hazardous wastes are greater than previously reported" [13, p. x]. The study found that the majority of people of color in the United States live within 3 km of a hazardous waste site. They are also more likely to have multiple hazardous waste sites in their communities. These same communities are then more likely to face the environmental risks of dumping such as air pollution, heavy metal contamination, and water quality degradation. They are also more likely to suffer from the health and psychological issues that arise from environmental degradation. Government and workplace regulations are least likely to be enforced in people of color neighborhoods [12]. Complaints of environmental poisoning from these neighborhoods are ignored not only by the regulatory agencies but also mainstream environmental organizations.

The story of the Holt family in Dickson, Tennessee is a striking example of this racism. The family lived 500 feet from the county landfill for decades. For part of that time, the carcinogenic, TCE or trichloroethylene, was dumped at the landfill site eventually leaking into the Holt's well water [20]. However, the family was never informed by the government about the TCE in their water. Government records show that officials were aware of the potential for TCE in the Holt's well but their well was left untested for nine years while the nearby wells of white families were tested. When testing on the Holt well finally revealed high TCE levels, an EPA official concluded that use of the well water would not harm the family's health. Only when several members of the family started suffering (and dying) from various forms of cancer did any questions come up. Even then, the family had to take the lead in investigating and suing the county [16].

This pattern of discrimination where communities of color and Indigenous communities disproportionately bear the environmental, health, financial, and political costs of environmental degradation is called environmental racism [61]. To be clear, environmental racism does not mean that white communities are not affected by environmental degradation. It does mean that when, for example, landfills are placed in white neighborhoods, local politicans are quick to address environmental concerns [12]. In those cases, landfill sites may be modified or even completely moved from the neighborhood. However, in neighborhoods with more people of color, those same landfill concerns may not be addressed. This environmental racism is part of a broader pattern of environmental injustice in North America that places poor and working-class neighborhoods, Indigenous peoples, and women on the losing side of decision-making and standards around waste [14].

2.2 ENVIRONMENTAL AND SOCIAL INJUSTICE

The environmental justice movement sees social injustice as being connected with environmental injustice [62]. The exploitation of women, people of color, Indigenous peoples, and the environment is caused by the same structures of oppression. The fact that marginalized groups disproportionately bear the burden of our waste mirrors the reality they face in finding housing, employment, and fulfilling other needs. Dismantling the common structures of oppression in parallel is necessary for both social and environmental justice. For example, engineers cannot talk about the costs of environmental degradation on different communities without addressing racism or class. To seek environmental justice then means dismantling the systems of capitalism, industrialization, racism, and consumption that have led to environmental degradation. In this chapter, examples of these connections are pointed out using the lens of social and environmental justice on waste.

2.2.1   CLASS AND WORK

Low-income communities are more susceptible to environmental harm [40]. When environmental decisions such as the placement of landfills are made, low-income communities are seen as "paths of least resistance" compared to higher-income communities. Not only do they usually lack the resources and political power to advocate for themselves, but they are also ignored by most environmental advocates. This means poor communities end up with lower air/water/land quality in their environment. However, poor people (especially poor people of color) are not just exposed to environmental degradation as residents, they are also exposed to it as workers [48]. For example, migrant farm workers are exposed to pesticides, laborers in steel factories are exposed to carcinogens, and production workers in Silicon Valley are exposed to toxics in electronics. The safety and environmental quality of these workers is not valued in their workplaces. Most have no means of seeking compensation for the effects of work on their well-being and health. Even if they do, many cannot risk losing their jobs or incurring the ill will of their employer as they are dependent on the job for income.

2.2.2   WOMEN AND THEIR FAMILIES

Mexican women in East Los Angeles fed up with the state of California's long-standing practice of placing toxic waste facilities in low-income or racialized neighborhoods organized into the Mothers of East Los Angeles (MELA) [47]. Their organization played an important role in opposing a proposed waste incineration plant in the town of Vernon near East Los Angeles [49]. The incinerator was going to be constructed within 7,500 feet of homes, schools, churches, hospitals, and food processing facilities. As part of its byproducts, the facility would release ash, dust, and heated gases containing carcinogens and mutagens. MELA opposed construction of the incinerator citing health and environmental concerns. They took the lead in opposing construction of the incinerator through months of legal battles, protests, and petitions until construction was cancelled in 1990. In being interviewed about their activism, members of MELA cited their roles as mothers as one of the primary reason for their actions against the plant [47]. Many felt that as protectors of their children, they had the responsibility to ensure that future generations would not inherit a degrading environment. These women organized to protect their community precisely because their families were being threatened by the actions of state of California, which saw their communities and families as paths of least resistance to building environmentally destructive projects.

2.2.3   INDIGENOUS PEOPLES, NIMBYISM, AND SOVEREIGNTY

The lands[1] of Indigenous peoples in North America have been used for the testing of nuclear weapons and the dumping of nuclear waste [15]. This environmental racism extends beyond nuclear waste though. Indigenous peoples are routinely on the losing side of environmental decisions because their sovereignty is unrecognized and they are the victims of NIMBYism[2] [42]. Abrogation of treaties with

---

[1]By lands, I mean the lands given to Indigenous peoples by settler states and not their original lands whether ceded or unceded.
[2]NIMBY=Not In My Back Yard

Indigenous people deprives them of sovereignty in making decisions about waste disposal on their lands. Their self-autonomy is especially important because NIMBY sentiments in white, higher-income, or settler communities forces environmental decision-makers to place waste on Indigenous lands, where there will be less resistance.

Waste injustice is a complex problem. It is not just an issue of corporations or the state versus the people. It is a historical issue, a political issue, and an economic issue that pits communities against each other. Sometimes it evens pits white environmentalists against Indigenous people bringing up questions of sovereignty, self-determination, and structural racism. We do have alternatives, though, to how waste is currently handled. Waste management is a social process that has been shaped gradually over time [52]. Engineers and other waste decision-makers can reshape the way the costs and benefits of waste handling are currently distributed. Taking the lead from the environmental justice movement [1], I propose that engineers analyze and make decisions around waste using the lens of social justice.

2.3 WHAT EXACTLY IS SOCIAL JUSTICE?

Social justice is a complex and fuzzy term. As Riley writes, this fuzziness is part of social justice [58]. Social justice can be defined in various ways. It is, however, always defined by a community, describes a process rather than an end, and is always changing as context changes. Examples of social justice include redistributing goods over a society to ensure a more equitable distribution, increasing protection for workers, prioritising environmental concerns over corporate profits, or resisting the oppression of people of color. However, there are many more ways to articulate social justice. For example, Jessica Yee connects justice for Indigenous youth with sexual education that promotes "for all sexes and genders, autonomy, personal responsibility, full and active consent, safety, health, happiness, pleasure, self-esteem, non-subordination and non-violence and true equality" [60, p. 26]. Yee views such education as fundamental to the stopping of sexual assault and gendered violence for future generations because it empowers youth (both Indigenous and non-Indigenous).

Different schools of thought also give their own "spin" to social justice. Faith-based understandings of social justice are different from anti-capitalist or critical understandings of justice although they may share common values. One example of an anti-capitalist and critical understanding of justice can be seen in the work of educator Paulo Freire. Freire focuses on how educators can employ liberative pedagogies to critique the existing oppressive worldview of their students and themselves [59]. He envisions a classroom where students and educators work together to liberate each other from the cycle of oppression. Through this conscience-raising, educators, and students can work towards social justice in themselves and their communities.

Given the fuzziness of social justice, Riley writes that "it is important, when working in a specific context, to be clear about what one means by social justice" [58, p. 27]. In this section, I define what social justice means for me in the context of this chapter and the following one.

I draw from the work of Iris Marion Young in writing about social justice. Young declares that justice is not just about fair distribution of goods but whether people are liberated or inhibited by social institutions [56]. She is responding to the traditional conception of justice in the humanities as equitable redistribution of goods [9].

This general conception of justice as redistribution has been articulated by many scholars each with their own idiosyncrasies and definitions [17]. The common thread in justice as redistribution is that society as it functions does not ensure goods are distributed equitably. Therefore, goods need to be redistributed to those who are marginalized. Debates may be had over what goods are, how they may be redistributed, why, and to whom, but redistribution is necessary for justice.

The continued marginalization of people of color, queer people,[3] women, Indigenous people, and other identity-marginalized communities combined with the rise of neoliberalism in the 1990s painfully highlighted the shortcomings of the framework of redistribution [33]. Justice as the fair redistribution of goods did not fully address the experiences of members in those groups, as their claims of oppression went beyond distributive concerns [32]. For example, Indigenous people demanded special recognition for the injustices they suffered in settler societies rather than merely sharing more fairly in the benefits of those societies.

To provide a theoretical foundation for those demands, Honneth proposed a theory of justice grounded in the struggle for recognition [37]. In Honneth's framework, social injustice occurs when recognition that is held to be legitimate by those demanding it is withheld. The withholding of recognition harms those who are not recognized by restricting their freedom to act and move towards greater self-realization [38]. Justice is then the struggle for recognition.

Young adopts both conceptions of social justice (as redistribution and as recognition) in her work. For her, justice not only refers to a more equitable distribution, "but also to the institutional conditions necessary for the development and exercise of individual capacities and collective communication and cooperation" [56, p. 39]. Injustice, on the other hand, is a result of forms of oppression that place structural constraints on the self-development and self-determination of particular groups of people because of the "unconscious assumptions and reactions of well-meaning people in ordinary interactions, media, and cultural stereotypes, and structural features of bureaucratic hierarchies and market mechanisms, in short, the normal processes of everyday life" [56, p. 41]. In defining injustice in this structural manner, Young draws from the struggles of feminists [8], anti-racists [22], queer people [54], Indigenous people [45], and other oppressed groups [2].

Young outlines five forms of oppression: exploitation, marginalization, powerlessness, cultural imperialism, and violence. Exploitation, marginalization, and powerlessness "all refer to relations of power and oppression that occur by virtue of the social division of labor" (p. 58). Cultural imperialism and violence refer to relations of power and oppression that occur through "cultural images, stereotypes, and the mundane reproduction of relations of dominance and aversion in the gestures of everyday life" [56, p. 63]. Central to Young's conception of justice are relations of dominating power and how those relationships allow some groups to have power over other groups. For example, Young links the domination of nature with the oppression of women, workers, and Indigenous people [55]. According to her, the same relations of power exploit the environment as well as marginalize people.

---

[3]See [10] for the context of why I use the word "queer".

In connecting justice with oppression and dominant power, I believe Young develops a powerful model for understanding justice. She is not without her critics, though.

Allen finds two important points are missing in Young's model: psychological oppression and power not just as oppression [5]. Allen writes that psychological oppression, the process by which the oppressed internalize self-images created by oppressors [7], should be another form of oppression in addition to the five already articulated by Young. In addition, Young thinks of power only as an oppressive force whereas Allen points out that power through resistance also plays a role in the collective and individual empowerment of the oppressed [5]. She argues that Young minimizes the role of resistance to dominant power in individual/collective empowerment by proceeding directly to participatory radical democracy as a solution to injustice. I agree with Allen on both points.

In using Young's conception of justice, I add Allen's view that resistance to dominant power is an important part of justice, so power can both be oppressive and empowering. Why I do this becomes clear later in this chapter, where I discuss some of the literature defining power and resistance. As a result, I propose the following as the working definition of justice for this book and as a framework for Chapter 3:

Working for social justice means to challenge forms of structural oppression such as exploitation, marginalization, powerlessness, violence, cultural imperialism, and psychological oppression through resistance to dominant relationships of power.

To be clear, this definition of justice is one rooted in a recognition-driven, liberal-social-democratic framework. The goal of resistance in this context is for oppressed groups to be recognized by a dominant order. That order may remain unchanged for the most part even after previously unrecognized groups are included. One critique against this framework is that a recognition-driven resistance does nothing to meaningfully challenge the power relations on which the dominant order is built. Many people, including anarchists [39] and autonomy-oriented Indigenous peoples [42], do not want to be recognized by an order that is oppressive in the first place. They want to walk away from the dominant order. I choose to define social justice within a recognition-driven framework for this chapter because I believe waste management decisions in North America currently take place within the context of liberal-social-democratic states.

2.4   POWER AND RESISTANCE

Power and resistance are two central concepts in the approach to social justice taken in this book. Essentially, social justice is resistance to the dominant relationships of power that underly structural oppression. Without understanding these relationships of power, I believe engineers cannot actively resist injustice through their work. What is power though? What is resistance?

These concepts deserve more of an explanation due to their importance. In the following sections, I discuss literature on power and resistance that can help engineers better understand those two concepts as well as their relationship to my definition of justice.

Humans, as individual and collective actors, can hold and exercise social power. Scott defines power generally as the capacity of an autonomous principal actor to create an intended causal effect upon a free subordinate actor possibly by restricting the choices available to subaltern actors [50]. There are five components to Scott's definition of power.

1. Power is, at root, the ability of one agent to intentionally affect the actions of another agent through their social relations. To take an example from my former academic institution, Queen's University, a group of white male engineering students exercised their social power over a female faculty member of color by pushing her off the sidewalk before insulting her with racial slurs [21]. In this case, the intended effect of the principal actors (the male engineering students) was to intentionally humiliate the subaltern actor (the female faculty member). While there was physical power involved in the interaction, there was also the social power that enabled the male students to humiliate.

2. The positions of principal and subaltern actors change in every context and are fluid even within a particular context. Although in the incident at Queen's, the students were the principal actors and the faculty member was a subaltern actor that necessarily would not hold true in other contexts. For example, in the context of a classroom, the students might be subaltern actors while the faculty member teaching the course would be a principal actor.

3. In every context, the principal and subaltern actors are free to choose from a range of options constrained by their context and relationships.

4. Without this freedom to choose, power cannot be exercised by principal actors as they cannot intentionally choose to exert power nor can subaltern actors choose to react to that power in resistance.

5. Finally, power can not only be exercised but also held. To continue using the example of the incident of racism at Queen's, the male engineering students have social power over females of color at Queen's due to the prevalent culture of whiteness even if they choose not to exert that power [41]. Therefore, power can be thought of as capacity.

   - While Scott provides a definition of power useful in introducing the concept, the term power actually has contested meanings within and between different theoretical philosophies. Haugaard points out that academic literature on social power can be divided into four theoretical categories: analytical political theory, nonanalytical political theory, social theory-modern, and social theory-postmodern [36].

Not only is power discussed differently in each theoretical category, but there are contests over the meaning of power within the categories.

Scott provides a simpler picture of academic research on power by dividing up the literature on power into mainstream and second stream research [50]. Mainstream research sees power as a zero-sum struggle where autonomous principals exert power to make subordinate agents comply to their needs while those subordinate agents resist by attempting to take power for themselves. In this view of power, there are clear losers and winners as principals and subordinate agents fight to claim a set amount of power in primacy. Academics writing in the mainstream of power research concentrate on studying specific organizations of power: the nation-state, multinational corporations, etc.

In contrast, the second stream of power research concentrates on the "strategies and techniques of power" [50, p. 9]. Second stream power researchers see power diffused throughout society in every social relation between every actor. It is not just limited to use by principals over subordinate agents. Second stream researchers are, therefore, interested in how power is deployed and used in every interaction throughout society. In addition, power is not just used by actors to repress. It may also be used productively to resist dominant power.

2.5 FOUCAULT'S FRAMEWORK

One of the most influential second stream theorists of power is Foucault [35]. Foucault's analysis of power as an intimate and localized force in everyday relations has been very useful in uncovering the dynamics of power, resistance, knowledge production, and truth-telling in asylums [31], prisons [30], and even on sexual bodies [28]. I believe it will be just as illuminating in uncovering the power dynamics of engineering projects. It is precisely because of Foucault's insistence on his theory of power not being a normative theory, but rather a way of uncovering power relations that makes it relevant to socially just engineering design [27].

A universal theory of power cannot explain how power, oppression, and resistance uniquely play out in every engineering design process without trying to fit the process into its framework. Instead, I take the position that engineers need a framework of analysis that can unpack the unique power dynamics of each design process.

Foucault's work on power provides a framework to unpack power dynamics. Using them, engineers can identify dominant (oppressive) power, relationships amongst actors, and shifts in relationships of power due to changing contexts and resistance. Foucault's work on power is also relevant for engineers because of his view of power relationships as a network between actors as opposed to the mainstream view of power as a one-line relationship between the principal and subordinate. While engineers usually work for an institution (the state, a private firm, a consortium, etc.), they must deal with many different stakeholders (neighborhood associations, city governments, community members, Indigenous representatives, etc.) all holding and acting upon the project and the engineer through different power relationships. I believe Foucault's "networked" view of power is the best approach for analyzing these diverse stakeholders relationships. In the following sections, I attempt to review Foucault's complex work on power and resistance with the hope that engineers will use his work to identify and resist the dominant relationships of power that underly structural oppression.

2.5.1 WHAT IS POWER?

Foucault sees power as both a productive and repressive force that acts to structure the field of possible actions for subjects of that force [26]. Unlike classical theorists of power, he does not think of power as an enforcement mechanism to control subjects. Rather, power conditions the types of actions the subject can take by structuring the field of possible actions in a certain way. For example, a subject of the Canadian state can choose to not pay taxes every year but then their field of possible actions is influenced by the laws of the state. They cannot just declare nonallegiance to the Canadian state (and independence from its laws) as that is not an option after refusing to pay taxes. In this case, the state through its various intermediaries will either compel them directly to pay taxes or shape their actions by changing the choices available to them (unemployment insurance, healthcare, etc.).

Power need not always repress a subject's possible actions. Power can also be productive by producing pleasure, creating knowledge, or establishing truths that shape a subject's field of actions [25]. Unlike violence which just seeks to repress a subject's freedom, power attempts to shape the subject's conduct by inciting, inducing, and seducing the subject to take particular actions [26].

2.5.2   WHERE IS POWER?

For Foucault, power is not a monolithic, singular, and periodic one-way projection of force from sovereign organizations or from principals to subordinates. Instead, power is diffused everywhere and comes from everywhere [32]. Power both conditions networks of relations and also operates through these same networks. Rather than being periodically projected, power constitutes all actors and relations at all times. No one and nothing is outside of power. In other words, power is a relationship between actors. It should not be confused with capacity, the ability of one actor to have power over another, or communication, transmission of information using language, even if all three types of relationship influence each other [26]. Even an actor like the state is not a monolithic entity. The state is actually a matrix of power relations between individuals that attempts to shape them as subjects.

2.5.3   HOW IS POWER NEGOTIATED?

Power is a network of relationships between all actors at all times. As these networks change in their actors and relations over time, place, and circumstance, the relations of power also change. In other words, power circulates through networks of actors constantly being renegotiated [25]. As new power relations are created through resistance or general power relations are extended, networks shift thus shifting power relations on other subjects as well. In addition, the constant negotiation of subjects through the field of possible actions conditioned by power in unpredictable ways forces power relations to shift. All this negotiation means that power is never evenly distributed. It is placing some subjects in better positions to negotiate power while putting other subjects in less favorable positions.

2.5.4   HOW DOES POWER OPERATE?

In his work, Foucault describes the techniques, technologies, apparatuses, and institutions used to exert power [25]. One major way power shapes the possible actions of subjects is by dictating discourses, or ways of thinking and producing meaning [43]. Discourses can include social practices, ways of creating knowledge, specific languages for a topic, and common-sense truths. At their most basic level, discourses influence how a particular topic is to be talked about at that historical moment while not allowing room for that same topic to be discussed in other ways. In fact, ways of talking about a topic not included in discourses at that historical moment can be said to not exist [24].

Another way power operates is through the idea of a rights-based framework of governance [25]. According to Foucault, the discourse of rights is an instrument of domination since it governs how subjects may have rights, which subjects have rights, and how those rights may be lost or gained. Subjects do not naturally have these rights. They are given to them in a particular historical moment and can be taken away under different circumstances.

Foucault provides a good example of power operating to affect behavior in his analysis of the modern European penal system. He argued that a new type of power called discipline has either replaced or works alongside rights-based power [30]. Disciplinary power normalizes individuals through material coercions and education using a system of surveillance. Since individuals know that they are under surveillance at all times (whether they are actually under surveillance is not important), they must normalize their behavior to meet the demands of those surveilling. The knowledge of surveillance itself is a part of the social reproduction process and couched in terms of service to the larger body [25]. While disciplinary power allows for a small number of specialists to condition a larger population, power is still diffused everywhere. The subjects of disciplinary power can restructure the field of power through their actions by making creative choices among the options they do have. In a prison, this might mean working on a prison farm or staging riots.

Power even acts upon bodies by changing and training them to become sources of labor [30]. Think about assembly line workers who have been trained to perform one task on the assembly line and no other task. They have been compelled by management to perform a task which has no meaning outside of the factory. In this example, training is a tool of power turns bodies into technologies of power.

2.5.5 WHAT DOES RESISTANCE HAVE TO DO WITH POWER?

In *The Subject and Power*, Foucault states "power is exercised only over free subjects, and only insofar as they are free" [26, p. 790]. The freedom that Foucault refers to is the agency of a subject of power to choose resistance to that power [25]. He argues that power, whether productive or repressive, and resistance are intricately connected in that resistance towards power must exist for power to be exerted and power must be exerted for resistance to occur. At the same time, power and resistance are mutually exclusive in that power attempts to remove the freedom of subjects to choose resistance. This means that power and resistance are always confronting each other, but also that power/resistance are two halves of a whole.

The intimate relationship of power and resistance does not mean that resistance is merely a reaction to power or a fight for a reversal of power relations. If power is seen as actions that restructure the field of actions for a subject, then resistance is the possibility that the subject can move through the field of possible actions in unpredictable and creative ways [26]. By choosing to navigate the field of actions in unpredictable ways, subjects challenge the ability of power to limit them. Resistance then is the means through which subjects can attempt to reshape some of the power relations exerted on them.

I adopt Foucault's views on power/resistance not just because of his rejection of a normative theory but also because he ties freedom (the possibility of restructuring the field of power) to resistance (a key component of my definition of social justice).

2.5.6   WHAT DO TRUTH/KNOWLEDGE HAVE TO DO WITH POWER?

Francis Bacon famously writes "knowledge is power" reflecting his belief that positivist scientific knowledge granted power [6]. Foucault turns this simple aphorism on its head by arguing that it is not knowledge that grants power, but power that shapes regimes of truth and the nature of knowledge [25]. As power legitimizes, shapes, and privileges certain truths and knowledges over others, those bodies of discourse, in turn, reinforce that power by normalizing particular behaviors and restricting others. Over time, these regimes of truth and knowledge become institutionalized, professionalized, and backed up by reward systems. Other types of discourses are marginalized.

For example, the medical sciences have been able to prescribe "normal" sexual behavior on a societal scale by building knowledge and truth around sexual behavior where previously those knowledges were fluid changing with time, place, and circumstance [32]. Meanwhile, non-desired sexual behavior has been explored by these medical sciences as a deviant practice. Over time, the medical sciences have come to be seen by other actors as a legitimate productive (creating not repressing) objective truth. They fail to see medical science as an extension of power that seeks to limit sexual behavior by uncovering and categorizing deviant sexual behavior. In short, medical science gets to establish the objective truth of sexuality.

The connected relationship between power/truth/knowledge does not mean that current regimes of truth/knowledge are indefinitely extended through the production of supporting discourses. The discourses can be contested by subjects as they resist dominating power [25]. One way to contest dominating discourses is through uncovering subjugated knowledges [29]. Subjugated knowledges are anti-structural, anti-monolithic, anti-hierarchical, and anti-science (non-institutionalized) knowledges. An example of a subjugated knowledge would be oral histories passed down among families, clans, ethnic groups. These histories are ignored in favor of standardized universal histories. Oral historiesare knowledges of local resistance to power that contest the totalitarian discourses put forward by power. Dominant discourses of knowledge marginalize subjugated knowledges by claiming that they are inadequate, naive, or not scientific enough [29]. However, it is precisely these qualities that make subjugated knowledges useful in challenging the totality of normalized discourses.

2.5.7   HOW TO ANALYZE POWER?

Can power be "fairly or objectively" analyzed if it permeates all knowledge, all relations, and all analysis? Would not an analysis of power itself be shaped by power? Foucault answers those questions in the positive by admitting that nothing, including analysis of power, can be outside of power.

Instead of trying to lift an analysis of power above society, he creatively takes the opposite route by embracing the role of power within an analysis of power. He proposes using points of resistance as an opening to identify and uncover power relations [26]. The struggle between the exertion of power and resistance towards that power becomes a way to uncover what power is being exerted, through what mechanisms, on which subjects, etc. For example, if we want to know more about power relations around engineering expertise, we should look at the relation between non-expert and expert knowledge.

Once power relations are uncovered, Foucault recommends asking five questions about those relations [26].

1. What differences in status and privilege allow certain actors, institutions, and subjects to limit the field of actions for other subjects? Are they differences of class, gender, race, legitimized knowledges, etc.? And how do those power relations create those differences while those differences, in turn, reinforce those power relations?

2. What are the objectives of actors exerting power over subjects in that particular situation? Is it, for instance, the accumulation of capital or the maintenance of white privilege?

3. Through what mechanisms is power exerted? Is it through military control, advertisement, discipline?

4. How are the power relations institutionalized? Is it through legal codes, knowledge regimes, or the family?

5. How effective are the mechanisms of power in achieving the objectives of the actors using those mechanisms both in terms of certainty of results and cost (whether in terms of economic or human cost)?

2.6 MODELING A SOCIALLY JUST APPROACH TO WASTE

Now that I have outlined my definition of social justice and provided some tools of analysis for power and resistance, how can we apply it to making decisions about waste?

Working from the definition of justice articulated earlier, I propose, as an answer to that question, that engineers should challenge structural oppression through using design techniques that resist dominant power relations. Remember that, according to Foucault's analysis of power, resistance is used by actors to reshape power relationships in their network. This means both power being exerted on them and power being exerted by them. Dominant power relations can be reshaped through resistance therefore changing or removing structural oppression. Through this resistance, engineers can individually and collectively empower people to challenge their oppression.

I believe one way to practically model engineers challenging dominant power is by looking at how other professionals, such as media activists, work for justice. Uzelman compares social justice oriented media activists to bamboo gardeners [53]. In a bamboo garden, single shoots of bamboo above the ground are actually the product of an intensive underground "complex network of root-like stems and filaments called a rhizome" [53, p. 17]. Before those shoots of bamboo even become visible, the rhizomatic network beneath the ground has to grow horizontally, establish new connections, and make space for the eventual garden.

Similar to the bamboo garden, social movements, fluid communities resisting dominant power, grow underground establishing connections with each other before sending out visible shoots of resistance. These social movements behave like the rhizome of a bamboo garden as they intersect with each other to fight different but intersecting forms of oppressive power. Uzelman argues that it is among these social movements that media activists work to not only present alternative views to mainstream media but also democratize media creation. This work takes place in the rhizomatic networks of social movements as media activists build alliances between communities of resistance, create democratic media spaces, resist dominant media power, and flatten hierarchies of media creation. Like the stalks in the bamboo garden, demonstrations and protests are the just the visible shoots of the nurturing done by media activists within social movements.

In the same way, engineers interested in justice must work in the rhizomatic networks of social movements to support and build resistance to oppressive design methods. This means those engineers would not unilaterally design technology or lead resistance in a community. They would rather be part of larger social movements in communities to resist dominant forms of power. While media activists create alternative media, engineers would work in the rhizome creating space for participatory design processes, flattening hierarchies of decision-making about technology, establishing alliances between those oppressed by existing technology, and designing technologies using methods that resist dominant power relations. The produced technologies would be the bamboo shoots of all the work by these engineers within social movements to create "just design."

Using the bamboo garden model of media activists, and the understanding of just design as resistance to dominant power, I propose a four-step approach to managing waste in a socially just manner.

1. Map relationships of power between actors.

2. Identify the structures of oppression created by and reinforcing the dominant relationships of power.

3. Identify, support, and nurture social movements that resist those oppressive structures.

4. Work within these social movements to build socially just technology using design methods that resist dominant power relations.

As shown in Figure 2.1, the design approach starts with the engineer mapping out relationships of power between actors including themselves in a particular project or field of work. This mapping would take into account unequal or dominating power relations between actors (i.e., Do men have more power in design decisions?—Is there any space for children's input?). Not all of the actors in a project may be recognized by an engineer though. Some actors may not want to be recognized by an engineer. So I would add the caveat that the engineer should try to map as many actors demanding recognition as possible. The engineer should not map actors by themselves. The process of mapping should involve other stakeholders as well since some stakeholders would be visible to those stakeholders but not to the engineers. This is the step in my proposed design process where an engineer would use Foucault's understanding and analysis of power.

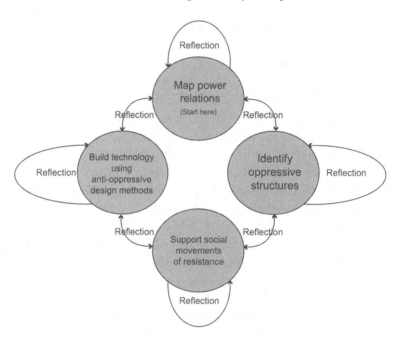

Figure 2.1: Proposed socially just design method.

In the next step, the engineer identifies the structures of oppression that are created by and reinforce the dominating power relations. (i.e., powerlessness or marginalization of women in design decisions). Again, not all structures may be recognized by the engineer, but the goal is to identify as many as possible.

In the third step, the engineer works within communities dedicated to resisting structures of oppression to build networks of solidarity, provide education on technical issues, and create space for equitable decision-making. Using the example of marginalization of women, an engineer may decide to ally with feminist and women's groups to ensure the voices of women are heard in design decisions about technology.

The final step for the engineer is to work within communities of resistance to build technology or make decisions using design methods that resist oppressive power relations. For instance, an engineer may use feminist design methods that resist patriarchal ways of designing [11] to ensure decisions about waste are just towards women.

The design process embraces reflectivity and nonlinear progress. It is reflective as engineers should take time to reflect on their role in the design process and the relations of power exercised by and on them. It is non-linear as the engineer can move forward, backward, or stay in the same stage of the design process at any time. Finally, the circular design process reflects the assertion by having no final step that relations of power cannot be absent. There is always the need to continually examine how power relations influence engineering design. Even if a design process is equitably just towards women, it may favor a certain subset of women such as upper/middle-class women or women racialized as white.

2.7   COMING BACK TO ENVIRONMENTAL RACISM

I started this chapter with a discussion of environmental racism, a type of structural oppression that falls under a broader pattern of environmental injustice. Under the definition of social justice, I proposed, for this and the following chapter, that environmental racism would fall under marginalization, psychological oppression, and even violence (violation of bodies and health). Voices in communities that deal with environmental racism are routinely marginalized from environmental decision-making and wider public discourse. Many in those communities must also deal with the mental and physical effects of environmental degradation on their psyches, families, and bodies.

For engineers interested in resisting such environmental racism, Foucault can help to understand what kinds of power relations are at play in their particular context. They can identify dominant power relations being exerted on communities of color to be docile and accepting of environmental decisions. They can also identify how the very same communities reproduce those dominant power relations and what kind of power relationships do they have internally (gender dynamics, class differences, etc.). Foucault's work on power can help engineers understand how environmental racism is established by existing social relations but also reproduced through those relations.

Finally, by using the proposed model of socially just design in Figure 2.1, engineers could understand how to resist environmental racism in its various forms.

To be clear, the design model that I propose does not just apply to resisting environmental racism nor do I believe that environmental racism is the only or primary cause of environmental injustice. I only use environmental racism as one example of injustice. Colonialism, class differences, and gender also factor into environmental injustice. The model that I propose could also be used to work with low-income or working-class communities, women, Indigenous peoples, and other marginalized communities to push back against multiple and overlapping forms of environmental injustice.

In the following chapter, I use my definition of social justice, Foucault's toolbox of power, and the proposed socially just design model to explore a landfill project in Canada on Indigenous land in much more detail.

REFERENCES

[1] David Naguib Pellow, "Environmental Racism: Inequality in a Toxic World." The Blackwell Companion to Social Inequalities, Mary Romero and Eric Margolis, eds. Blackwell Publishing, 2005. 12

[2] Maurianne Adams, Warren J. Blumenfeld, Rosie Castaneda, Heather W. Hackman, Madeline L. Peters, and Ximena Zuniga. Readings for Diversity and Social Justice. Routledge, New York,US, 2000. 13

[3] Jeffrey C. Alexander and Maria Pia Lara. Honneth's new critical theory of recognition, November-December 1996 1996.

[4] Taiaiake Alfred and Jeff Corntassel. Being indigenous: Resurgences against contemporary colonialism. Government and Opposition, 40(4):597–614, 2005.
DOI: 10.1111/j.1477-7053.2005.00166.x

[5] Amy Allen. Power and the politics of difference: Oppression, empowerment, and transnational justice. Hypatia, 23(3):156–172, 2008. DOI: 10.1111/j.1527-2001.2008.tb01210.x 14

[6] Francis Bacon. Religious meditations of heresies. Essayes, religious meditations, places of perswasion and dissuasion. Etchells and H. Macdonald, London, 1st ed. of 1597 edition, 1924. 19

[7] Sandra Bartky. On Psychological Oppression, pages 33–41. Philosophy and Women. Wadsworth Publishing company, Belmont,US, 1979. 14

[8] Bell Hooks. Feminist theory: from margin to center. Pluto Press, London,UK, 2000. 13

[9] David Boucher and Paul Kelly. Social Justice: from Hume to Walzer. Routledge, New York,US, 1998. 13

[10] Robin Brontsema. A queer revolution: Reconceptualizing the debate over linguistic reclamation. Colorado Research in Linguistics, 17(1), June 2004. 13

[11] Cheryl Buckley. Made in patriarchy: Toward a feminist analysis of women and design. Design Issues, 3(2):3–14, Autumn 1986. 22

[12] Robert D. Bullard. Confronting environmental racism: voices from the grassroots. South End Press, 1993. 9, 10

[13] Robert D. Bullard, Paul Mohai, Robin Saha, and Beverly Wright. Toxic wastes and race at twenty years 1987–2007. Technical report, United Church of Christ, 2007. 9

[14] David Enrique Cuesta Camacho. Environmental injustices, political struggles: race, class, and the environment. Duke University Press, 1998. 10

[15] S. Chakma, M. Jensen, and International Work Group for Indigenous Affairs. Racism against indigenous peoples. IWGIA document. IWGIA, 2001. 11

[16] Michelle Chen. In our backyard: Environmental racism in dickson, September 2009. 10

[17] Matthew Clayton and Andrew Williams. Social Justice. Blackwell Publishing Inc., Malden, US, 2004. 13

[18] Kia Corthron. Splash hatch on the E going down. Dramatists Play Service, Inc., 2002. 9

[19] Angela Davis and Elizabeth Martinez. Coalition building among people of color. Inscriptions, 7:42–53, 1994.

[20] Lynne Duke. A well of pain. The Washington Post, 2007. 10

[21] Gloria Er-Chua. Queen's slow to act against racism: prof, January 11 2008. 15

[22] Joe R. Feagin. Systemic racism: a theory of oppression. Routledge, New York, US, 2006. 13

[23] Commission for Racial Justice. Toxic wastes and race in the United States. Technical report, United Church of Christ, 1987. 9

[24] M. Foucault. Archaeology of knowledge. Routledge, 2007. 2002067999. 18

[25] Michel Foucault. Power/knowledge: selected interviews and other writings, 1972–1977. Random House, Inc., New York, US, 1980. 17, 18, 19

[26] Michel Foucault. The subject and power. Critical Inquiry, 8(4):pp. 777–795, 1982. DOI: 10.1086/448181 16, 17, 18, 20

[27] Michel Foucault. The Foucault reader. Random House, Inc., 1984. 16

[28] Michel Foucault. The history of sexuality: An introduction, volume 1. Penguin Books, New York, US, 1984. 16

[29] Michel Foucault. Genealogy and social criticism, chapter 3, pages 39–45. The postmodern turn: new perspectives on social theory. Cambridge University Press, New York, US, 1994. 19

[30] Michel Foucault. Discipline and punish. Random House, Inc., New York, US, 2nd edition, 1995. 16, 18

[31] Michel Foucault. Madness and civilization: a history of insanity in the age of reason. Routledge Classics, Abingdon, UK, 2001. 16

[32] Michel Foucault. The History of Sexuality: The Will to Knowledge. Number v. 1. Penguin Group, 2008. 13, 17, 19

26 REFERENCES

[33] Nancy Fraser and Alex Honneth. Redistribution or recognition? : a political-philosophical exchange. Verso, New York,US, 2003. 13

[34] L. Gandhi. Postcolonial theory: a critical introduction. Columbia University Press, 1998. 97032402.

[35] Gary Gutting and Michel Foucault. The Stanford Encyclopedia of Philosophy. The Metaphysics Research Lab, Center for the Study of Language and Information, Stanford University, spring 2010 edition, 2010. 16

[36] Mark Haugaard. Power: a reader. Manchester University Press, New York,US, 2002. 15

[37] Alex Honneth. The struggle for recognition: the moral grammar of social conflicts. Polity Press, UK, 1995. 13

[38] Axel Honneth. Recognition and justice. Acta Sociologica, 47(4):351–364, December 01 2004. DOI: 10.1177/0001699304048668 13

[39] N. J. Jun and S. Wahl. New perspectives on anarchism. Lexington Books, 2010. 2009015304. 14

[40] J.P. Lester, D.W. Allen, and K.M. Hill. Environmental injustice in the United States: myths and realities. Westview Press, 2001. 11

[41] Natalia Maharaj. The experiences of racialized female faculty at queen's university, 2009. 15

[42] D. McGovern. The Campo Indian landfill war: the fight for gold in California's garbage. University of Oklahoma Press, 1995. 11, 14

[43] A. W. McHoul and W. Grace. A Foucault primer: discourse, power, and the subject. New York University Press, 1997. 17

[44] David Miller. Principles of Social Justice. Harvard University Press, Cambridge, US, 1999.

[45] Patricia Monture-Angus and Mary Ellen Turpel. Thunder in my soul: a Mohawk woman speaks. Fernwood Publishing, Halifax,Canada, 1995. 13

[46] R. C. Morris. Can the Subaltern Speak? Reflections on the History of an Idea. Columbia University Press, 2010.

[47] Mary Pardo. Mexican american women grassroots community activists: "mothers of east los angeles." Frontiers: A Journal of Women Studies, 11(1):pp. 1–7, 1990. DOI: 10.2307/3346696 11

[48] D.N. Pellow. Garbage wars: the struggle for environmental justice in Chicago. Urban and industrial environments. MIT Press, 2002. 11

[49] Joel Reynolds. Toppling a toxic incinerator. 11

[50] John Scott. Power. Polity Press, Malden,US, 2001. 15, 16

[51] Amartya Sen. The idea of justice. Harvard University Press, Cambridge,US, 2009.

[52] S. Strasser. Waste and want: a social history of trash. Henry Holt and Co., 2000. 12

[53] Scott Uzelman. Hard at Work in the Bamboo Garden: Media Activists and Social Movements, pages 16–29. Autonomous Media: Activating Resistance and Dissent. Cumulus Press, Montreal,Canada, 2005. 20, 21

[54] Michael Warner and Social Text Collective. Fear of a queer planet: queer politics and social theory. University of Minnesota Press, Minneapolis,US, 1993. 13

[55] Iris Marion Young. "Feminism and Ecology." special issue of heresies: A feminist journal of art and politics 4, no. 1 (1981). Environment Ethics, 5(2):173–179, 1983. 13

[56] Iris Marion Young. Justice and the Politics of Difference. Princeton University Press, Princeton,US, 1990. 13

[57] Iris Marion Young. Unruly categories: a critique of Nancy Fraser's dual systems theory, chapter 2, pages 50–67. Theorizing multiculturalism: a guide to the current debate. Blackwell Publishers Ltd., Malden,US, 1998.

[58] Donna Riley. *Engineering and Social Justice*. Morgan & Claypool, 2008. 12

[59] Paul Freire. *Pedagogy of the Oppressed*. The Continuum International Publishing Group, New York, US, 2006. 12

[60] Jessica Yee. Sustainable Justice Through Knowledge Transfer. Canadian Woman Studies, 28(1):22–27, 2009. 12

[61] David E. Newton. *Environmental Justice: a reference handbook*. ABC-CLIO, Denver, US, 2009. 10

[62] R. D. Sandler and P. C. Pezzullo. *Environmental justice and environmentalism: the social justice challenge to the environmental movement*. MIT Press, 2007. 10

CHAPTER 3

Expertise, Indigenous People, and the Site 41 Landfill

Usman Mushtaq

"There was never refuse like this on reservations," he told his seminar, because, he said, walking backward to the window, "on the old reservations the tribes were the refuse. We were the waste, solid and swill on the run, telling stories from a discarded culture to amuse the colonial refusers. Over here now, on the other end of the wasted world, we meditate in peace on this landfill reservation." (p. 101)

Gerald Robert Vizenor [124]

3.1 INTRODUCTION

In Chapter 2, some of the effects of environmental racism on communities of color in North America (Turtle Island[1]) were described. The effects of this racism extend to Indigenous[2] communities in North America as well. These communities disproportionately bear the costs of environmental pollution, negligence, and decision-making [132, 133, 134, 135]. This chapter shows some of the costs, focusing on an example of decision-making around siting of waste (Section 3.2).

The goal of the chapter is not only to show how Indigenous people are marginalized in waste siting and disposal, but also how engineers can work with Indigenous communities to decrease or remove the environmental and human costs of improper waste siting/disposal. I believe engineers have a crucial role in championing and including Indigenous communities in environmental decision-making. However, this can only happen if environmental decision-making is seen through a social

---

[1]Turtle Island refers to North America in the mythology of many Indigenous communities [131]. In this chapter, I use the term "North America" over Turtle Island to be consistent with the other chapters in this book and to make the material more accessible. I acknowledge that I have this privilege because I live and write as a settler on unceded Algonquin territory. When I use the term Canada, I use it to refer to the state of Canada as a political and government entity and not the land.

[2]I use the "term" Indigenous peoples to refer to the native peoples of the Americas. I do this over using the terms "Native American" or "Aboriginal" or "First Nation" to acknowledge the sovereignty of those peoples distinct from settler states imposed upon them, and to recognize peoples who are Indigenous to the Americas but have not been recognized as so by settler states. At the same time, I acknowledge the distinct and separate nature of all the Indigenous communities since they don't belong to the same Aboriginal "culture" [23]. I capitalize the first "i" in Indigenous to acknowledge that Indigenous people are not just an ethnic group but also sovereign political entities. If I was describing them as an ethnic group (Indigenous would be an adjective), but as a political entity "Indigenous" is a proper noun.

justice lens (Sections 3.3, 3.4, and 3.5). Only then can engineers make environmental decisions that champion marginalized voices (Section 3.6).

In the case of Indigenous people in North America, environmental injustices are connected to their lack of sovereignty and self-determination. Take the Oka Crisis in Quebec as an example. In the summer of 1990, the municipality of Oka, the Canadian state, and Mohawks were all pitted against each other over the use of Mohawk land called the Pines [2]. The Mohawks had ecologically stewarded the Pines since 1717, when they were "granted" the land by France after being removed from their traditional lands. Yet, the Oka municipality gave away this unceded[3] land to developers without any consultation with the Mohawk community. The crisis happened when the municipality unveiled plans for expansion of a golf course already on Mohawk land. The move happened without any consultation with the Mohawk community even though it threatened Pine Hill Cemetery, a small cemetery used by the Mohawks, and the Pines were unceded Mohawk territory. Not only did no consultation take place with the Mohawk community but no environmental impact assessments of the golf course expansion took place even though the expansion required massive deforestation, use of pesticides/fungicides, removal of animal habitats, and the paving over of the community's cemetery.[4] The Mohawk community took a stand against the loss and poisoning of their land, which eventually lead to armed conflict between the Canadian military and Mohawk warriors. In this case, the environmental injustice of cutting down forest land for golf courses was directly connected to the Mohawk's ability to steward the land.

This theme of environmental injustice against Indigenous peoples is repeated in various forms all across North America [35]. The Aamjiwnaang First Nation reserve located on the shores of the St. Clair River south of Sarnia, Ontario is surrounded by several petrochemical, polymer, and chemical industry plants leading to high rates of asthma and disproportionate births of girls [136]. The birth rate ratio was found to be 33% boys and 67% girls [137]. The First Nation has raised these concerns with the government but have so far been ignored. In fact, the government has even approved increased production rates for chemical facilities located in the area [138]. Since 1969, the community of Grassy Narrows located north of Kenora in Ontario has dealt with the mercury poisoning of their water through the effluent of a pulp mill [139]. The poisoning of their water has led to high concentrations of mercury in fish [140] as well as a variety of short-term and chronic illnesses in the members of the community [141]. The community has responded since then by resisting further exploitation of their land by the government and private companies through blockades, direct actions, and other forms of community organizing [142]. The remote Metis community[5] of Black Tickle, Labrador has limited access to water, most of it contaminated due to a lack of sewage treatment facilities [143]. This combined with the lack of facilities to test water safety make the community home to a high rate of illnesses. Indigenous peoples have similarly been on the "losing" side when it comes to siting of waste facilities by the Canadian government.

---

[3]Unceded meaning that the land had not been given to the Canadian state whether through treaty or force at any point in time.
[4]In contrast, the Osprey Ridge Golf Course Project in Nova Scotia (ACOA, Project # 05-94-0256) went through an environmental screening in 1996.
[5]Metis are Indigenous people who trace their descent from mixed European and Indigenous ancestry.

3.2   AKWESASNE AND GENERAL MOTORS

In 1958, General Motors opened a foundry near the Akwesasne reservation located across the U.S.-Canada border in New York State. This foundry ended up making the reservation one of the highest PCB-polluted sites in North America [58]. The foundry, along with the other factories built to support it, had dumped PCBs into designated waste sites near the reservation. However, these sites leaked the toxins into the surrounding ecology. When tests were conducted in the 1980s, PCB concentrations in fish and animals at Akwesasne were thousands of times higher than legally allowed. The contamination poisoned the land and water on which the Akwesasne people depended on for their food, economy, hunting, and fishing. A costly cleanup mandated and supported by the Environmental Protection Agency of the United States in 1990 has given too few resources to the Akwesasne in the cleanup of their land. Recent research highlights the continuing negative health impacts of the PCB contamination on all aspects of the Akwesasne community, especially the health of youth [112]. At no time was the community of Akwesasne consulted about the placement of the waste sites or their approval of the foundry. GM did not even follow-up on environmental testing of plant/animal life in the area of the waste sites and foundry.

When it comes to waste disposal, other Indigenous communities have similarly not been consulted or informed leading to irreparable poisoning of their health, land, and ecology. In general, Indigenous communities in North America have limited means to control environmental impacts on their lands [17]. The state may consult them on land use at which point the Indigenous community can state their opposition. However, the decision on final land use always lies with the state. Consultation is not enough for these communities since they do not often have the resources to adequately represent or fight for themselves in the state's legal systems. At the same time, the state brings pressure on these communities because they are seen as paths of least resistance compared to non-Indigenous communities when it comes to environmental exploitation. Indigenous communities cannot control environmental impacts for two reasons: a lack of Indigenous sovereignty and restrictive or vague avenues for state participation of the Indigenous nation. Historically, Indigenous communities have been left at the margins in policy development, which means the laws and policies governing their participation and consultation in the settler state are not always clear or followed. At the same time, they can do little to control impacts on their own land because, in many cases, unceded land is taken away from them or even if the land is under Indigenous ownership, a variety of state regulations take away self-determination from the communities. Without either the autonomy to self-determine land use or clear rights/procedures within the framework of the state (that the state is willing to enforce or observe), Indigenous communities are left with a restricted number of tools to fight back against environmental exploitation.

In this chapter, I explore the case of one landfill on Indigenous land using the social justice lens to show how, among other injustices, Indigenous peoples continue to be marginalized. At the same time, I apply the model developed in Chapter 2 to show how waste planning may be handled in a more just manner.

3.3   SITE 41

The story of North Simcoe Landfill Site, commonly known as Site 41 after its Ministry of the Environment (MOE) designation, begins in 1979 with the story of another landfill. The Pauze landfill, located near the vicinity of the future Site 41, was a privately owned 20 acre unlined landfill site that began serving North Simcoe County in northern Ontario in 1966 [21]. By 1979, the towns of North Simcoe[6] (under the North Simcoe Waste Management Association or NSWMA) began to see that the 20 acre site would not be sufficient for their waste needs. An engineering firm, Gartner Lee and Associates Limited, was hired by the towns to investigate a longer term waste solution [38]. One of the solutions investigated was an expansion of the Pauze landfill. However, evidence of illegal dumping of industrial wastes along with leachate seepage into the aquifer canceled any future plans for the landfill.

However, the county was not ready to start construction on another landfill site and no pre-existing landfills were sufficient for their needs. While they searched for a suitable site, the Pauze landfill kept being used as a landfill site until 1987 despite the environmental concerns already pointed out by the Gartner Lee and Associates Ltd investigation. This decision continues to have negative environmental and social implications. A 1992 assessment of the Pauze landfill by the MOE found the previously discovered leachate plume had not "undergone significant retardation within the aquifer" even after traveling a distance at which the "plume was predicted to become indistinguishable from background groundwater quality" (p. 51) [66]. The assessment recommended further monitoring to ensure that the aquifer in time would naturally attenuate the plume. However, even 18 years later, reports from well testing in the area indicate trace amounts of TCE and PCB toxic compounds that lead back to the Pauze landfill [76].

By 1986, Site 41 (Figure 3.1), near the town of Midland, had been selected as the landfill site for North Simcoe after the NSWMA went through an Environmental Assessment (EA) process [38]. The proposed placement of the site would impact two Indigenous communities: the Kawartha Nishnawbe First Nation and Beausoleil First Nation [72].

3.4   A HISTORY OF SITE 41

In this section, I present a history of Site 41 from 1986–2010 in order to provide a context for my later discussion of engineering design discourse at Site 41. As with any complex engineering project, a variety of voices, stakeholders, concerns, and histories were present during the Site 41 debate. I do my best to represent those concerns and histories. I do not claim the history I present is the only history of Site 41. Rather, I present a history of Site 41 primarily through sources that were marginalized in the Site 41 discussion (non-expert community members, farmers, women, Indigenous peoples, and citizen's groups). Even in 1984, during the initial site selection process by the NSWMA, these groups were being excluded from the discussion even though they would have to deal with the impacts of

---

[6]Midland, Penetanguishene, Tay, Port McNicoll, Victoria Harbour, and, originally, Tiny.

Figure 3.1: Location of the proposed Site 41 in Ontario (Source: Google Maps).

the waste site [72]. My goal is to make sure these groups are no longer excluded by pointing out how they were marginalized and how engineers can work with them towards justice.

Before the Site 41 selection could be finalized, a review of the EA process was undertaken by a Joint Board (a quasi-judicial government body with members selected from the Environmental Review Tribunal and Ontario Municipal Board [122]) as specified by The Consolidated Hearings Act [10]. This Joint Board ruled in 1989 that the NSWMA[7] in the EA process had a "predisposition to have site 41 selected as preferred - a predisposition that indicates bias" (p. 30) [38]. The Board further noted that the EA and site selection processes were flawed to the point of not being logical, traceable, or replicable. Therefore, they ruled that the EA process was deficient and the Site 41 landfill could not be constructed. The process was so fundamentally unjust that the Ontario Ministry of the Environment commissioned a report in 1990 to analyze the Joint Board decision on Site 41 [103]. The goal of the report was to draw lessons for future EAs from the failures of the Site 41 proponents to consult the public, do social impact assessments, present meaningful choices for landfill alternatives, or value the voices of non-expert community members.

---

[7]By 1989, the NSWMA did not include the Town of Tiny, which opposed Site 41.

Despite the history of the Pauze landfill and the failed Site 41 EA process, Ontario Premier David Peterson passed an Order in Council in 1990 rejecting the Joint Board decision by allowing the Site 41 proponents to reconsider the failed site once again as a potential landfill [30]. The NSWMA did not even have to conduct another EA of Site 41. They just had to present additional evidence of the technical suitability of Site 41, if available, to a second reconvening of the Joint Board. The suitability of the site with respect to the "agricultural community, or regarding conditions" did not have to be considered (p. 3) [30]. As support for the order, Council cited the testimony of a representative of the engineering firm (Jagger Hims Ltd) hired by the NSWMA to conduct hydrogeological studies of Site 41. According to the firm, the likelihood of finding a better site than Site 41 was remote even though the Joint Board had written that the site selection process was deficient. Over 40 community submission on other possible locations, waste solutions, or in support of the Joint Board's decision were ignored by the Council since they were not "qualified witnesses" (p. 2) [30].

From 1990–1993, eleven additional sites were investigated by the NSWMA and its consulting engineers but Site 41 was reselected as the final choice. Once it became clear in 1992 that Site 41 would be reselected, the WYE Citizens group applied for funding established by the Environmental Assessment Board to help Site 41 stakeholders make their respective cases [60]. The WYE Citizens group, originally consisting of 40 families living near Site 41, formed in 1985 during the first Site 41 EA to voice North Simcoe resident concerns. They were opposed to Site 41 during the initial Joint Board hearings citing the impact of the landfill on their properties, businesses, and lives. In fact, they played a central role in bringing up some of the failures of the first Site 41 EA to consider social, community, and agricultural impacts.

The funding request was successful, but the group was limited to using the funds to hire a technical representative [60]. Even though they had asked for funding for an expert on social/community impacts and the 1989 Joint Board hearings had shown the deficiency of the social/community aspects of site selection, the Environmental Assessment Board would not release funds for non-technical matters. Still, the WYE Citizens group did their best to voice their concerns during the second Joint Board Hearings on Site 41 from 1993–1994 [37].

In addition, the corporation of Tiny Township, where Site 41 would be located, took part in the hearings opposing Site 41 for many of the same reasons as the WYE Citizens group. Tiny even filed a legal challenge against the Joint Board questioning the Board's authority to resume hearings on Site 41. This challenge was denied and the hearings began in May of 1993.

In the hearings, the chair of the Joint Board, Eisen, continuously implied that he was "governed or perhaps more accurately constrained by the wording of the OIC [Order in Council]" (p. 16) [37]. Eisen admitted that the site selection process presented to the 1989 Joint Board was "virtually non-existent yet my interpretation of the OIC obliges me to accept the opposite to be the case" (p. 14) [37]. He goes on to say: In spite of the fact that no new evidence was introduced in connection with this matter, I must accept not only that such a process exists but further that it is logical, traceable and replicable. (p. 14) [37].

Furthermore, the constraints imposed by the OIC kept the Board from considering the social, community, and agricultural impacts of Site 41 despite the concerns raised by North Simcoe residents. Even when the Board found the public consultation efforts of NSWMA lacking "the type of flexibility and imagination required by the unique circumstances of these proceedings," Eisen was forced to acknowledge that they were adequate enough for the Order in Council. The lackluster public consultation effort was certainly not because of community disinterest. The WYE Citizens group presented a ten point waste management plan emphasizing recycling and reuse as an alternative to Site 41. The plan was rejected [104].

At the end of the hearings, the Joint Board determined that it was satisfied with the new technically sound site selection process and analysis, and so approved Site 41 with certain conditions in 1996. The Joint Board's decision was immediately challenged by Site 41 opponents in the Ontario Divisional Court and later appealed in the Ontario Court of Appeals. However, in March and July 1997, respectively, both the Ontario Divisional Court and the Ontario Court of Appeals dismissed the challenges [106]. In April 1998, the MOE issued a Certificate of Approval (C of A) along with conditions of approval for landfill construction at Site 41 [102]. The C of A gave Simcoe County[8] the authority to build a landfill if they submitted a Development and Operations report to both the MOE and the Community Monitoring Committee (CMC) for approval. The CMC was established by the C of A to provide "community review of the development, operation, ongoing monitoring, closure and post-closure care related to the landfill site" (p. 1) [34]. From this point on, the CMC group would be crucial in voicing concerns of justice for Simcoe residents and groups marginalized in the landfill development process.

Before Simcoe County could file a Development and Operations report, the C of A was challenged in front of before the MOE [90]. The challenge was based on a February 2001 Simcoe County Environmental Services Committee Report presenting new data that indicated that groundwater would enter the landfill site, making the use of a landfill liner mandatory. MOE denied the challenge, arguing the applicants had not demonstrated in their evidence that the landfill would cause significant social, economic, or environmental harm. Later, the Environmental Commissioner of Ontario[9] would state "MOEE's [Ministry of Environment and Energy] handling of this application was disappointing" (p. 227) [90]. Among other issues, he argued the MOE had not been transparent in its answer to the challenge nor did its adequately engage with the community's concerns over possible drinking water contamination.

In January 2003, Simcoe County submitted the Development and Operations Report prepared by engineering firm Henderson Paddon to the MOE [64]. The MOE responded in March 2003 with initial comments on gull management on site before responding with a full review in June 2003. The full MOE review identified 85 issues with the Design and Operations report [118]. Some of these issues were the impact of the landfill on a nearby creek, incomplete data on parts of the landfill

---

[8]In 1995, the NSWMA disbanded and Simcoe County took over as the Site 41 proponent [10].
[9]The Environmental Commissioner of Ontario is the environmental watchdog for the province [92].

site, and the need for pumping of groundwater to an offsite location. By this time, the concerns of the MOE dealt primarily with technical issues at Site 41.

In parallel, the CMC and Tiny Township commissioned independent peer reviews of the Development and Operations Report.[10] The peer reviews concluded that constructing a landfill on Site 41 would be possible; however, several issues would have to be addressed before construction could begin if the landfill was going to be environmentally sound. The Tiny Township peer review explained the site would "require that more than usual precautions are undertaken to design, construct, and operate the Site in accordance with the requirements" (p. 11) [34]. The problems of public consultation or social impact were not brought up in any of the reports since the consultants were not mandated to address those problems.

Based on the MOE comments and the peer reviews, citizen's groups filed yet another challenge in 2004 requesting a review of Site 41 by the MOE [106]. The MOE declined to undertake the review, writing that many of the issues had already been identified through public consultation and further issues might be resolved by Simcoe County in follow-up reports. The Environmental Commissioner of Ontario in his annual report remarked that this decision only focused on specific technical factors, which was "unfortunate" [91]. He also countered the claim of public consultation put forward by the MOE stating "that many of the technical details related to the landfill's design and operation were not available during the landfill siting process or at the hearing" (p. 153) [91]. Despite all the criticisms, the MOE granted approval to the Design & Operations Report for Site 41 on October 2006. Ray Millar, chair of the CMC at the time, claimed in a letter that "it became virtually impossible for the CMC to access information from Simcoe County, and it was equally apparent that the MOE was not supportive of further enquiry by the CMC" from this point on (p. 4) [75]. When the CMC asked for clarification of their appeal rights from the MOE, they were told by the senior Review Engineer that they no longer had any appeal rights left.

However, not every avenue of opposition had been exhausted. New evidence in support of stopping construction at Site 41 was published in 2005/2006. Testing of the groundwater at Site 41 by a group of scientists revealed extremely low concentrations of antimony [116]. Further testing confirmed that the groundwater contained only trace amounts of other metals [77]. Residents of north Simcoe County argued that water under Site 41, comparable in purity to arctic water samples, should be preserved [109]. The MOE and Simcoe County responded that the water did not meet drinking water guidelines in Ontario; however, Shotyk, the lead author on the Site 41 groundwater studies, pointed out the only parameters not met were aesthetic ones such as hardness and iron [55].

Shotyk would go on to publish further reports validating the natural filtration qualities of the Site 41 aquifer that produced water with trace amounts of metals and organic contaminants after the decision was made to go ahead with construction of the landfill [114, 115]. Activists against the Site 41 landfill expressed concerns that such a clean water source may become contaminated by leachate from the landfill.

---

[10]Golder Associates Ltd., Urban and Environmental Management Inc., and Dr. R.K. Rowe Inc. undertook the review for the CMC [16]. Dixon Hydrogeology Ltd., Conestoga Rovers and Associates Ltd., and Severn Sound Environmental Association undertook the review for Tiny Township [34].

Despite the evidence raised by the groundwater tests, Simcoe County Council voted in June 2007 in favor of allowing construction to start by a vote of 16 to 15.

Stephen R. Ogden, a member of the CMC, filed a request in September 2007 [45] with Simcoe County requesting a copy of the hydrogeological model created by Jagger Hims of Site 41 [45]. The model was created in Modflow, an open source software program. This hydrogeological model was the basis for the selection of Site 41 and the MOE approval. The CMC was interested in gaining access to the model created by the Modflow program so that it could be independently peer analyzed. The County denied the request in October 2007 because the model information was not under the control of the county but rather Jagger Hims [61]. At the time, the county made no effort to obtain the model from the engineering firm. The CMC appealed that decision to the Information and Privacy Commissioner of Ontario (IPC), still seeking to gain access to the model. The appeal succeeded and the IPC appointed a mediator to resolve the case. As part of this mediation process, representatives from the CMC, Simcoe County, and Jagger Hims met three times in September/October 2008 to review the Modflow model [63]. However, Jagger Hims would not publicly release the model nor their inputs into the model. Ray Millár, who was a CMC representative at the mediation meetings, said "the tripartite meetings did not answer his questions, nor the concerns raised by the CMC's consultant" [50]. He later added that the MOE had "failed to acknowledge their lack of expertise" (p. 5) since they had not analyzed the inputs of the Modflow model in deciding to issue the C of A [75]. Mediation had failed.

Meanwhile at Site 41, construction of the landfill infrastructure (roads, fences, ponds, etc.) had started in June 2008 [108]. The construction necessitated the removal of water flowing out of the site.[11] Simcoe County filed for a Permit To Take Water (PTTW) at the MOE to allow water from the landfill to be discharged after treatment [43]. Critics of the PTTW pointed to the large discrepancy between the water planned to be removed in the Development and Operations report and the PTTW [74], the effects of dewatering on the water table [81], and the impacts of discharging treated water in a local creek (MacDonald Creek) and its aquatic life [101] as evidence that the PTTW should not be granted. To address some of these concerns, the CMC requesting funding from Simcoe County to hire a consultant to peer review the PTTW [70]. The funding was denied but representatives from Jagger Hims were made available to CMC to answer questions about the PTTW [27]. The Department of Fisheries and Ocean, after a petition by Steven Ogden (a CMC member), also submitted comments on the potential impacts of draining water into MacDonald Creek [28]. A monitoring program was proposed to track the impact of water being discharged into the creek. Still, the CMC felt there were too many unanswered questions. They recommended that the PTTW not be approved until those questions could be answered [74]. The MOE granted Simcoe County a PTTW on December 8, 2008 [108].

Seeking additional avenues of support, Mohawk elder/activist Danny Beaton, along with Stephen Ogden, and other activists, opposed to Site 41 walked from the landfill site to Queen's

---

[11]Groundwater flows up and out of Site 41 because it is at an upward hydraulic gradient. Therefore, water needed to be removed from the site for construction [51].

Park[12] to raise awareness of the water issues at Site 41 [44]. On November 21 2008, they arrived at Queen's Park along with 300 supporters [14]. At this point in the Site 41 discussion, the landfill was no longer a local issue due to the PTTW filed by Simcoe County with the MOE along with the groundwater test results. These two issues thrust Site 41 into the national spotlight. The march to Queen's Park and the appeal to the Ontario government garnered media attention while the filing of PTTW made landfill opponents believe the landfill siting process was not fair. Groups such as the Council of Canadians, Sierra Club of Ontario, and the Green Party saw Site 41 as an example of freshwater mismanagement [82]. They argued that some of the cleanest water found in Ontario should not placed under a landfill noting the need to properly manage dwindling freshwater resources.

Maude Barlow, national Chair of the Council of Canadians, led a march in May 2009 to protest construction of Site 41 along with over 500 people [47]. This "Walk for Water" was followed by another one in July 2009 that attracted over 700 people [87]. Larger citizen's groups like the Council of Canadians were able to bring in more resources against Site 41 such as organizing these "Walks for Water" or even threatening lawsuits against Simcoe County [83]. However, some of the fiercest and most meaningful opposition to Site 41 came from local farmers and Indigenous activists, and in particular the Anishinabe Kweag (women from the Beausoleil First Nation). On May 8, 2009, five women from the Beausoleil Nation set up a permanent protest camp (Figure 3.2) across the Site 41 construction field [22]. Their goal was to protect the water of the Alliston Aquifer, of which Site 41 was a part, for future generations.

The peaceful camp remained for 137 days serving as a site of protest and support for both Indigenous and non-Indigenous opponents of Site 41. On June 12, 2009, protesters at the camp, frustrated by the Site 41 process, nonviolently blocked the entrance of Site 41 preventing construction staff or equipment from entering the site [119]. The Anishinabe Kweag took the lead in leading the blockade at the site.

Indigenous groups also took the lead in other ways around the Site 41 protests. The Metis Nation of Ontario came out against Site 41 reminding the government that they have a "legal obligation to consult and accommodate Metis on activities that have the potential to adversely affect Aboriginal rights or interests" (p. 1) [12]. That had not been done during the 30-year history of Site 41. Chief Kris Nahrgang representing the Kawartha Nishnawbe First Nation expressed similar concerns in a letter to Simcoe County Council letting them know that the nation had not been notified of Site 41, let alone consulted, even though the site was on their treaty and traditional hunting/fishing land [80]. Chief Rodney Monague Jr. of the Beausoleil First Nation addressed a letter to Simcoe County asking for a moratorium on construction because of the possible contamination effects of the landfill on "all the water within our territories" [144].

Protesters continued to block entrance into Site 41, leading Simcoe County to file an injunction against the protesters citing the financial cost of delayed construction [78]. On July 22, 2009, the Ontario Superior Court ruled Simcoe County had the authority to bar and remove protesters from

---

[12]Queen's Park is the site of the Ontario Legislature and the Government of Ontario.

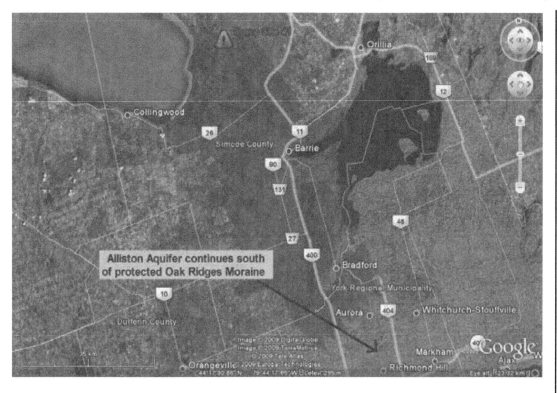

Figure 3.2: Site 41 location above Alliston aquifer highlighted in blue [7].

Site 41 who were blocking the entrances. After a written warning to the protesters [48], Ontario Provincial Police (OPP) charged members of the protest camp with mischief, carried out raids on the camp, and arrested some protesters [46]. They disrupted the protest site by going in to remove tents and signs [24]. Other protesters were ordered to report to police precincts in order to be charged and arrested for what Canadian Union of Public Employees President Sid Ryan said was their political dissent [13]. The arrested protesters were barred from the site until the conclusion of their court trials in December 2009 [85], which all ended in charges being stayed under the condition that none of the protesters would be involved in civil unrest at Site 41 for the next year [49].

At the same time, the IPC assigned an adjudicator to resolve the ongoing Modflow case between Ogden and Simcoe County. The adjudicator, after an investigation, issued an order to Simcoe County in May 2009 to direct Jagger Hims to submit the Modflow model and inputs to the records of the county [18]. In June 2009, Simcoe County issued a request to Jagger Hims for the Modflow model [62]. Jagger Hims responded they that it would not release the Modflow model as the model was Jagger Hims' intellectual property [15]. The adjudicator responded with an order to

Figure 3.3: Part of the protest camp—Mohawk flags, a symbol of Indigenous resistance in Canada, were prominent [86].

Simcoe County to take all actions, including legal ones, to obtain the Modflow model from Jagger Hims [19].

Simcoe County responded to the IPC that the county had no position with regards to the Modflow model and did not know of any legal or other types of actions they could take to obtain the model [52]. However, still under pressure from the IPC, they continued to negotiate with Genivar [13] until a settlement was reached in November 2009 [53]. The settlement required Genivar to provide the Modflow model to pre-approved peer reviewers for analysis while still retaining ownership of the model. The model would not be publicly released and the engineering firm would have a chance

---

[13] In 2009, Jagger Hims Ltd. was bought by Genivar Consulting Inc., a major landfill developer, who had advocated for Site 41 in front of Simcoe County Council in 2007. Genivar had also bought Henderson Paddon, the engineering firm responsible for Site 41 development, earlier in 2008 [84].

Figure 3.4: Rally for a moratorium on Site 41 [86].

to respond to any of the peer reviewer reports before public release. Ogden declined to accept a settlement that "purports to bind him without his prior involvement, and that assumes his eventual approval of the terms of the agreement" (p. 1) [11]. On February 12, 2010, the IPC stayed both previous orders (MO-2416 and MO-2449) because Simcoe County filed requests for judicial review of the orders [20].

All the concerns, unheard voices, and missed opportunities for consultation finally led over 2,500 people to attend a rally on July 25, 2009 in Perkinsfield (Figure 3.4) to demand a moratorium on Site 41 construction from Simcoe County [8].

On August 25, 2009, Simcoe County councillors, facing pressure from national and local citizen's groups, Indigenous communities, local farmers, and Simcoe County residents, voted 22-10 in favor of a one-year moratorium on construction at Site 41 [88]. This vote was followed by another vote in September halting construction of the Site 41 landfill permanently without revoking the C of A [89]. A private member's bill introduced by Simcoe North MPP Garfield Dunlop sought to revoke the C of A but it failed to pass [107]. Simcoe County Council requested revocation of the C of A to the MOE on May 25, 2010 [36] and the MOE complied shortly [42]. Site 41 was rezoned by Simcoe County Council as an agricultural property with the condition that it could not be used for waste management purposes.

3.5   SEEING SITE 41 THROUGH A SOCIAL JUSTICE LENS

In the last chapter, working for social justice was defined as challenging dominant power relations. So seeing Site 41 through a social justice lens means that power relations around decision-making at the landfill need to be uncovered. One way to do that is through the use of Foucauldian Discourse Analysis (FDA).

3.5.1   DISCOURSE ANALYSIS

FDA practitioners believe that language produces and reproduces power relations in two ways: by constraining what can be said, by whom, where, and when [31] and by constructing objects that become objective through the use of particular vocabulary [129]. In other words, language (or discourse) creates our reality for us by limiting how we talk about certain subjects and who has more authority to talk about them. For example, in the field of mental illness, schizophrenia has an objective meaning that is used by medical professionals. There are unwritten rules around how schizophrenia is used (in a hospital) and by whom (psychologists). Through this, schizophrenia has become an objective way to describe a particular form of mental illness that excludes other ways of describing that form of mental illness. This also means that treatments for schizophrenia are restricted by what is accepted or considered objective. FDA practitioners would argue that the language used to describe the form of mental illness known as schizophrenia is constructed because of the particular power relations in North American society that privilege "medical" ways of talking about illness.

Power is expressed through words and text. Language masks unequal power relations by setting dominant discourse as a given, as the "way it is" between different social groups (men and women, rich and poor, or colonial settlers and Indigenous peoples). The goal of FDA, then, is to reveal the rules and relations of power that govern dominant discourse. In other words, it is to reveal what is "given" within a particular truth. This is done not to get to the objective truth (according to Foucault there is no truth beyond discourse [59]), but to uncover the means through which power privileges certain versions of truth.

While FDA shares many of the principles of broader discourse analysis (i.e., discourse being constructed by and constructing social reality) [59], there are some themes unique to FDA. The first one, already discussed above, is the intimate relationship between language and power. Dominant power not only works through discourse but also constructs it [129]. Going back to the example of mental illness, odd behavior is generally discussed in terms of schizophrenia as opposed to witchcraft, because medical science dominates the discourse around odd behavior. Medical science dominates other ways of talking about health at this particular point of history. This makes medical science a given as opposed to witchcraft although at other times in history in various cultures, witchcraft has been a given. The dominance of medical science in describing mental illness happens not only through the language of medical professionalism but also by setting the terms and participants of the discourse around odd behavior. A spiritual healer's views could not be included in the discourse around odd behavior since they not seen as objective. A second theme, also mentioned above, is that there is no correct version of the truth. Since Foucault argues that all truths are constructed by power relations, it would be impossible to arrive at a power-free truth. Instead, Foucault tries to show through his analysis how "the what, whom, and why have been constituted in discourse" (p. 156) [129]. This allows FDA practitioners to reveal the link between power relations and dominant discourse.

Discourse can be an important part of resistance to dominant power. Established discourses can be disrupted through the creation of alternate discourses [127]. Alternate resisting discourses "can be used to challenge power, to subvert it, to alter distributions of power in the short and long term" (p. 11) [128]. Resisting discourses are often created by actors from oppressed social groups in the form of a counter-culture. Practically, zines, demonstrations, blogs, social media, or grassroots co-op media may comprise the texts of the counter-culture [123]. An example of such a counter-culture in the discourse of mental health is the efforts of the Icarus Project. The project contends disrupts the discourse of mental illness put forward by medical science in favor of a community wellness approach that sees experiences labeled "mental illness" as "mad gifts needing cultivation and care, rather than diseases or disorders" [145].

In these situations of contention between discourses, texts serve as archives of struggle showing the negotiations of power that took place in creating that text [128]. By texts, discourse analysts mean any speech, written document, or visual image. Texts serve as archives of struggle through details such as the placement of words/images in a text, the method of delivery of a speech, symbolism invoked by the use of particular words, etc. [129]. Through studying such details, FDA practitioners can determine the influence different social groups had over the text. For example, by looking at health policy around mental illness, FDA practitioners may be able to discover what power relations constructed the text, what kinds of discourses are negotiated, and what discourses are marginalized in the language used and who is allowed to speak in the text.

In particular, I will be using Foucault's version of discourse analysis termed archaeology to analyze the discourses in texts on Site 41 such as government documents, news reports, and stake-holder interviews. Foucault defines archaeology as a method to reveal discourses, the unverbalized rules behind those discourses, and how those rules govern discourse in distinct intellectual fields [39]. The rules that govern discourse are the implicit assumptions that shape what can be said about a subject and by whom. Using archaeology on Site 41, texts will reveal those implicit assumptions in the language around Site 41 and how those given assumptions shape what is said about the waste site.

The basic unit of analysis in archaeology is discursive formations. Discursive formations are groups of statements, objects, or concepts subject to specific rules of formation. The rules of formation are the conditions of meaningful existence for a discursive formation maintained by conditions of power. For example, the use of the pronoun "he" to refer to males or the pronoun "she" to refer to females is an unwritten rule of formation that defines gender in English-speaking society into two distinct binary genders. The assumptions in English are that "he" cannot be used to refer to females (or vice-versa) and that all people must fit into the "he/she" gender binary. This leaves no way for people to talk about themselves using fluid gender language, which reinforces the dominance of the gender binary. The rules that define use of "he/she" are not formal rules written down somewhere. The rules of formation vary from being general rules that apply to all the elements in a discursive formation to more specific rules. The goal of archaeological analysis is to analyze these rules of formation for their connections to power as they traverse a discourse from its most general form to more localized extensions.

1. Analysis of these rules of formation is always localized to particular discursive formations.

2. Analysis is localized to a particular domain because the aim of archaeology is not to create totalizing histories or identify universal rules of discursive formation. Instead, archaeology concentrates on describing the particular continuities, transformations, and contradictions of a specific domain such as the use of the "he/she" gender binary in North America. A domain is simply the field in which particular subjects are described showing similarities/differences at the level of rules of formation.

3. Archaeological analysis shows how different statements, objects, concepts, and other elements of a discourse may be formed by the same rules of formation, how these rules may or may not apply in the same way, and how similar/dissimilar concepts or objects in different discourses may actually come to existence respectively, through different/similar rules of formation. For example, many of the same rules of formation that apply to the use of "he/she" also apply to how subjects may dress. Male-identifying people may not wear skirts but female-identifying people can wear skirts, revealing the relationships between discourse and non-discursive domains like institutions, political events, etc. [39].

Discourse analysis should reveal how particular discursive formations are created and used by institutions or other distinct bodies while, in turn, supporting those bodies. To take the example of the male/female gender binary again, the government may provide different social and health services based on the gender binary that those with more fluid gender roles may not fit into. At the same time, the use of the gender binary allows the government to categorize people into clear gender roles (females are not allowed to serve in certain government positions, for example).

3.5.2   ANALYZING A DISCURSIVE FORMATION

If the basic unit of archaeological analysis is the discursive formation, then statements are the building blocks of discursive formations. By statement, Foucault does not mean a linguistic unit like a sentence. A statement is, rather, a function that relates discursive units to other units. It places discursive units in a "domain of coordination and coexistence" (p. 106), so that each statement is bordered by other statements [39]. The statement is not visible like a linguistic unit, but it is not hidden either. The relations it builds can be identified, so the statement can be known in its function. A statement also possesses a material existence in that it has different roles depending on the medium of discourse. When analyzing discursive formations, statements are not identified for their content but for their conditions of coexistence, functions, relationships to other statements, and material coordinates. As part of this analysis, statements may be interrogated for authorship (i.e., what is the author's position?), the site of the authorship (i.e., where is the author speaking from—a position of medical authority or one of a patient?), and the relationship of the subject to the domain of discourse (i.e., medical science can speak with authority about mental illness but no so much about cars). An example of statements within a text could be the mental health conditions described in the Diagnostic and Statistical Manual of Mental Disorders (DSM) put out by the American Psychiatric Association. The DSM sets standards around mental health for psychiatrists around the world. The ways in which mental health conditions are defined and related to each other and to symptoms/treatments define the "conditions of coexistence, functions, and relationships" of mental health in medicine.

Another type of discursive unit that can be analyzed is the object. Objects arise as singular concepts in existing intellectual fields (i.e., madness in the mental health field). They are designated and formed by authorities of delimitation, who have the authority to use and restructure the particular intellectual field (i.e., doctors in the field of mental health). The description of objects takes place using forms of specification, which classify and describe the object. Objects do not exist until they are designated and described in their fields. To go back to the example of the DSM, schizophrenia as an object does not exist in medicine until it is defined by the writers of the DSM. That does not mean the conditions that schizophrenia describes do not exist but that it may be described or understood differently (through the lens of spirituality for example). For this reason, objects are defined for the purpose of archaeological analysis by external relations not by their internal definitions. Archaeological analysis on schizophrenia should take place by comparing understandings of the conditions that schizophrenia describes in other domains.

While objects are defined in existing intellectual fields, concepts are a sub-unit of discursive formation that emerge and circulate among a field of statements. They build upon other groups of statements in a succession. In analysis, these concepts can be divided into three categories: field of presence (group of statements accepted by a specific discourse), field of concomitant (group of foundational statements), and field of memory (group of statements no longer accepted but were precursors to current concepts). Concepts may be changed through procedures of intervention that are dictated by discourse. Analysis of concepts look not at the succession of concepts but at the rules of formation that make concepts possible, the organizing principles that move concepts from one type to another, and the procedures of intervention. An example of a concept in medicine is that mental illness can be diagnosed through a reading of symptoms. The concept of diagnosis is a foundation of medicine and the statements that comprise the concept are a field of concomitant. Without the foundation of diagnosis, there is no reason to categorize, map, and establish relationships between symptoms, illness, and treatment.

Finally, statements, objects, and concepts are all organized into broader themes that Foucault calls strategies. Strategies can be analyzed at points of conflict, where different strategies try to occupy the same discursive space. The actual conflict can be analyzed but also the rules that determine the principles of exclusion and inclusion for a strategy being challenged. Analysis of these rules should necessarily include the authority that sets rules to determine exclusion/inclusion of a conflicting strategy. These contradictions within discursive formations, instead of being ignored as aberrant, should be used for archaeological analysis, since they are the site at which power and discourse is being negotiated. An example of a contradiction in mental health would be the exclusion or inclusion of certain items in the DSM from year to year. Why were these items included or excluded over time?

3.6   EXCAVATING POWER AT SITE 41

Several subjects emerged from my analysis of engineering design texts at Site 41. In particular, the texts spoke of the following subjects in particular ways: landfills, the environment, consultation, expertise, uncertainty, local knowledge, social impacts, authority, community resistance, and Indigenous peoples. The rules of formation that governed how these subjects were talked about defined and were defined by the relationships of power at Site 41. The texts I used for analysis came from the following sources:

- Ministry of Environment documents;

- Simcoe County reports & meeting minutes;

- StopSite41 web articles;

- local news media;

- interviews of Simcoe residents/Site 41 activists;

- Council of Canadians documents/website;

- technical reports from Jagger Hims/Genivar;

- peer reviews of Site 41 designs; and

- speeches by Simcoe County elected officials.

3.6.1 LANDFILLS

The Site 41 landfill was always presented as a necessity because no alternatives were viable. Landfills were "just the reality" because waste is always produced. For example, Simcoe County noted that the landfill is needed because everyone produces garbage [95]. That everyone produces garbage and that a landfill is the only way to deal with garbage were given assumptions. The necessity of landfills precluded any discussion of meaningful alternatives to the landfill [54]. Even when talking about alternatives such as waste diversion, the landfill was always argued to be needed [93]. When the Joint Board received an application from the WYE Citizens group discussing landfill alternatives, it praised their submission but noted that there is still a need for the Site 41 landfill [37]. This discourse of waste as a given and landfills as a necessity served to limit the design decisions around Site 41. This directly affected consultation as many of the suggestions by the residents of Simcoe County were ignored since they did not acknowledge the necessity of landfills. Therefore, it fell to the technical experts to decide on how that waste is to be handled: through a landfill.

Reinforcing the theme of landfills as necessary was the strategy of presenting increasing waste and decreasing waste handling capacity as a crisis [32]. Waste, continually being produced, is piling up and if Site 41 is not approved, it will have negative impacts on other landfills, on people, on the environment, on waste collection, etc. [95]. In particular, the negative impact on other landfills if Site 41 was not constructed was cited by Simcoe County in pushing the urgency of the project [98]. The strategy of waste as a crisis played into devaluing community consultation and giving power to engineering experts in deciding how to deal with waste.

Landfills were also spoken of in terms of safety. The vocabulary of impacts, regulation, compliance, and soundness was used to point out how a necessary technology like landfills are safe. In particular, authority was invoked in expressing claims about safety. If the MOE has found Site 41 scientifically sound for safe use by the County [120], then what objections could there be to the site? Speaking of Site 41 in terms of safety served to discredit opponents to Site 41 since their concerns had nothing to do with safety (or at least safety using the vocabulary of bureaucracy and science). In a CMC meeting, Douglas Jagger used that vocabulary (impacts, sensitivity, mitigation) in discussing concerns the committee had about Site 41 [27]. The monopoly of experts can be seen in this meeting as a vocabulary of science and bureaucracy is deployed.

3.6.2 ENVIRONMENT

Site 41 design texts discussed the environment primarily in terms of protection. The environment had no agency or role beyond being protected. Even in discussions of protection, the environment was a passive subject kept safe by human-engineered features [68]. These engineered features would be built by experts after conducting appropriate studies [99]. Statements about the spiritual value of the ecosystem or protection of the "lifeblood of Mother Earth" made by Site 41 opponents had no meaning in Site 41 design [126]. Both the Jagger Hims consultants [57] and Simcoe County [98] adopted the discursive strategy of talking about the environment in terms of protection. Engineering expertise thus defined the conditions and vocabulary under which the environment could be understood. When this discursive strategy was challenged by Site 41 opponents, design texts highlighted the natural protective features of Site 41 in addition to the human engineered features [100]. The natural protection was highlighted to show the appropriateness of Site 41 even without the need for human engineering features. However, the environment was again discussed only as a means to protect itself from the landfill [126].

Within this discursive strategy, protection as an object was defined in three ways. One way was to talk about protection as safeguarding human health and safety. This approach to environment protection centered around humans and their use of the environment [121]. When the MOE representative discussed the PTTW with CMC members and engineering consultants, terms like fair, sustainable, and efficient use were highlighted in the discussion [105]. Protection of the environment had the assumption of human use built into it, implying the environment did not have intrinsic value beyond its use by humans.

Other definitions used in talking about protection were protection as management, and protection as compliance. The Joint Board in 1989 heard from engineering consultants that impacts at Site 41 could be mitigated through proper management [38]. In conducting their peer review of Site 41, engineering consultants for Tiny Township also noted that careful management of the site would protect the environment [34]. Again, the environment is a passive subject that needs to be carefully managed. In addition, management implicitly ties environment protection to expertise through the deployment of a vocabulary of monitoring (sensitivity, limits, thresholds, etc.). The assumption is that non-experts, who do not have access to such vocabulary, cannot properly manage Site 41. In this way, community members were distanced from design decisions about how to protect the land.

Much of the vocabulary of monitoring came from compliance to MOE rules [56]. This is not surprising since Site 41 designers talked about protection as complying with MOE procedures, industry standards, or legislative codes. These codes and standards were brought up whenever site designers were challenged by non-experts. However, this only served to further distance non-experts by deploying a vocabulary of standards. With this vocabulary, engineers were able to ally themselves to institutional authorities such as the MOE. Of course, the environment as an active agent is absent from all this discussion of compliance. MOE rules and legislative codes determine whether the impacts on gulls [16] or water flow [100] need to be considered, not community members or the needs of the animal/plant ecosystem itself.

3.6.3   CONSULTATION

Consultation in Site 41 design texts was framed as convincing the community of the landfill. Consultation did not mean participation. Representatives from Jagger Hims viewed the public meetings as just a venue to present results or answer questions [57]. The aim of consultation was presented as minimizing confrontation between experts and non-experts [38] by eliminating misunderstandings held by the non-experts [38]. When concerns of source water protection were brought up by the CMC, the public was shown illustrations of the engineered protections at Site 41 after the experts had already decided on the design. They explained to the public why the protections would work.

Even the information sessions were designed in the words of one moderator to convey technical information effectively and efficiently [38]. This discursive strategy only alienated non-experts from the debate by limiting their role to passive recipients of information rather than contributors to design.

When engineers did talk about public input in design, they talked about addressing public perceptions and attitudes [57]. This language of perceptions and attitudes masked another discursive strategy used by Site 41 designers. Members of the public participating in consultation were shown to be emotional and acting on misinformed perceptions [67]. Site 41 designers showed public members to be irrational by pointing out their unanimous opposition to the landfill even in the face of expert advice supporting the landfill [57]. Engineers highlighted the emotions of public members at consultations to explain away this behavior. In contrast, the experts were implied to be objective and in control [103]. Objectivity, in particular, was seen to be more important than emotion, which is part of the overarching discourse of engineering. So even though the language of consultation was used, it was actually being used to bar non-experts from participation. Engineers used their time at the consultation meetings to sit in front of the audience answering questions instead of engaging with public concerns.

Consultation was also talked about as a requirement being fulfilled. Technical experts were accommodating the demands of the public in setting up meetings with them or sending information to the CMC [57]. This accommodation took place because of consultation requirements for the Joint Board [37] or the MOE [105]. The requirements to consult with the public did not come about organically or as part of the larger process. They were not seen by the Site 41 designers to have any valuable meaning attached to them. Rather, the requirements to consult were seen as a box to be checked off. In this way, designers prioritized the values of those in authority without consulting the public. When the CMC asked for funding from Simcoe County to hire a consultant to review a landfill base preparation test report in 2007, the County denied the request using the language of fulfilling their requirements of consultation [69].

Even when opponents of Site 41 challenged the experts, their concerns were only addressed if they came from peer review by other experts. The CMC commissioned several engineering firms to review the reports produced by Simcoe County's consultants. The concerns raised by these reviews were not described using the language of emotions or misinformation. Site 41 designers pointed to this as good consultation in contrast to the bad consultation where non-experts raised concerns out of misinformation or emotions.

3.6.4   EXPERTISE

The selection of Site 41, according to Simcoe County and its experts, was not only the result of an expert-driven process but also just "common sense" [32]. Expertise, backed by engineering jargon, was equated with "common sense" in the discussions around Site 41. For example, in selecting Site 41, the engineers had made "optimal" decisions based on a variety of factors unlike the emotions behind decisions made by non-experts [57]. This vocabulary of optimization (factors, variables, best fit lines, etc.) was deployed to prove the objectivity of experts. However, it only discredited the concerns of non-experts and community members by linking their objections to the landfill to emotions and perceptions. Non-experts and community members were portrayed as having no common sense.

In addition, Site 41 designers continually emphasized that engineering professionals (the experts) could be trusted to protect the environment and ensure the safety of humans. They could be trusted because of the "solid" science [95], independent facts and know-how [15], and expert training [100] behind them. Trust, as an object, was based on expert privilege and science knowledge, not community consultation or local knowledge. In fact, this trust was used to justify Site 41 construction even when there was fierce opposition to the landfill. In other words, trust was employed in silencing Site 41 critics through the use of experts [54].

Backing up this construction of trust was a discursive strategy that separated expertise from values. Experts could be trusted because their decisions were objective. While opposition to Site 41 may be driven by politics, its selection as a landfill only took place after extensive evaluations and studies [73]. Jagger Hims consultants argued that their focus is just on the technical (objective) aspects of landfill design [57] as did other experts during the Joint Hearings [37]. In fact, the MOE was quick to point out that the Joint Board reminded experts that they should not be advocates, only objective experts, in their study of the Board's 1989 decision [103]. By positioning expertise as value-less, the landfill and the decisions that led to the construction of the landfill were also positioned as objective. Not only did this strategy reinforce the emotional nature of non-expert concerns, but it also allowed engineers to distance themselves from community concerns. They could fall back on defending the site through asserting the scientific soundness of the site or the compliance of the site with standards.

Within this discursive strategy, technical vocabulary (engineering jargon) became associated with objectivity, and local vocabulary/concerns were associated with the political. When Gerretson, the MOE Minister, was questioned about the Site 41 debate, he stated the site was scientifically sound but that the MOE was ready to revoke the C of A if local authorities made that decision [121]. He implied that the call to cancel Site 41 was a local (political) decision not based on the facts of scientific soundness. During the Site 41 debate, engineers always presented their technical reports as objective facts, never acknowledging that engineering jargon has its own politics in how it devalues emotions, values human-centered understandings of the environment, and promotes particular methods of measurement. Most of the second Joint Board Hearings in 1995 were a debate between experts on hydrogeology or land units in technical jargon [37]. The public and their concerns were invisible during the debate excluded from the discussion. Engineers reinforced their monopoly on facts by only responding to community concerns if those concerns came in the form of engineering peer reviews [105]. Other local concerns were positioned as political issues that engineers could not address as objective experts. They could only consider issues which they considered to be engineering-related, limiting the breadth of concerns that were addressed.

3.6.5   UNCERTAINTY

Jagger Hims explained to the second Joint Board on Site 41 that it had only looked at major trends and conditions in siting the landfill in North Simcoe [37]. It argued that a more in-depth investigation would not be possible because of the complex geography of the region. This was one of the few times when the complexity and uncertainty of designing Site 41 was explicitly brought up by an expert.

Site 41 engineers used modeling as a way to cloak uncertainty in their decisions without acknowledging the inherent uncertainty of natural environments (for example in the flow of groundwater) [118]. They were able to explain away gaps in their knowledge through using modeling [111]. When non-experts challenged the predicted effects of climate change on hydraulic conditions at Site 41, representatives from Jagger Hims wrote to the MOE stating that their modeling of the site had already produced the likely effects of climate change [56]. The discursive strategy of using modeling to mask uncertainty worked to reinforce the monopoly of experts at Site 41 since only they had access to the model and the knowledge it produced. Criticism of the model was also seen by the Site 41 designers as a political or emotional act. Modeling also reduced the environment from a living ecosystem to a set of inputs/outputs. The environment was, again, a passive agent that was being modeled.

Objects such as risk and significance were constructed as part of this discursive strategy. In the first [38] and second [37] Joint Board decisions on Site 41, the Board was satisfied with engineered protections since they would minimize risk. The unpredictability of those protections was never explicitly discussed in the hearings though. Newsletters from Simcoe County about Site 41 emphasized that dewatering would not significantly affect wells in the area [93]. Again, though, the uncertainty of that knowledge was not acknowledged, just cloaked in the vocabulary of significance. Even experts not directly involved in the Site 41 engineering debate talked about uncertainty in terms of risk by arguing that risk at the landfill could be monitored and dealt with in the future [32].

3.6.6   LOCAL KNOWLEDGE

Local knowledge was invisible for the most part in Site 41 design texts. None of the engineering reports produced by Simcoe County's consultants made any explicit distinctions between local/non-local knowledge. However, most of the knowledge included in the reports came from modeling and environmental testing. Knowledge held by local residents about their environment was never obtained through interviewing or surveying them. The lack of local knowledge was reflected in CMC meetings, where non-expert locals found mistakes made by the engineers in water volumes [29] and creek flow rates [25].

When local knowledge was talked about, it was equated with the political and emotional. Jagger Hims engineers framed the local concerns that formed the basis for opposition to Site 41 as public outcry and pressure [57]. They equated local knowledge with the emotional, therefore, making it subjective as opposed to the objective knowledge they held. Local knowledge was made circumspect because of its emotional content. However, local knowledge was also circumspect because of its political content. In the earlier section on expertise, I wrote about how the local became political. In the same way, local knowledge was political and, therefore, subjective knowledge even if it used the same vocabulary/discourse as non-local engineering knowledge. When Simcoe County landfill experts talked about the results of water testing by Bill Shotyk, they were sure to mention that he was a local when presenting his lab data [100]. In the case of Shotyk's findings, his local-ness made his expert findings suspect.

3.6.7 SOCIAL IMPACTS

The first Joint Board on Site 41 was concerned about the social impacts of Site 41; however, their concerns were framed around the loss of productivity and business risk [38]. They only paid attention to social impacts such as loss of productive farmland or risk to nearby businesses (airport, farms, etc.) not to the emotional or spiritual impacts of the site. When the MOE analyzed the Board's decision a few years later, the study highlighted the same deficiencies in social impact assessment, but, again, only focused on business and working arrangements [103]. Other types of social impact were not invoked.

3.6.8 AUTHORITY

Engineers evoked authority in three ways while designing Site 41: authority as a means for proving their claims of expertise, authority as a means for defending their design decisions, and authority as a means to discredit opposition to the landfill.

Throughout the design of Site 41, designers pointed out that the MOE and Joint Board were both giving their approval to the design process [57]. Engineering consultants for Simcoe County would bring up the presence of MOE inspectors as proof that they were following proper procedure [65]. A vocabulary of legislative requirements, compliance standards, and environmental regulations was deployed in talking about how engineering design at Site 41 was meeting insitutional approval [68]. By associating their expertise with institutional authority, engineers placed their claims over the claims of community members.

Authority was also a means for defending design decisions at Site 41. If engineering decisions were challenged, engineers could fall back to the language of requirements. When Site 41 opponents brought up non-technical concerns, they were ignored as their concerns did not fit any of the design requirements set by the MOE. Even at the second Joint Board hearings [37], community concerns were ignored as the OIC had removed the requirements for social impact or community health assessment. This meant that opponents of the landfill were limited to critiquing only certain parts of the project in certain ways as determined by institutional authority. As long as engineers met technical requirements, their design decisions could not be critiqued.

This drove opponents of Site 41 to use more direct forms of resistance such as the protest camp and the blockade. Site 41 designers responded by portraying the protesters as criminals and troublemakers by invoking the authority of the state. In their reports to the County, engineers from the Henderson Paddon firm pointed out the costs of hiring site security to protect the landfill from vandalism [65]. When Simcoe County held a Site 41 information session in 2009, the Ontario Provincial Police (OPP) was patroling visibly in police cars around the meeting [5]. This used authority to discredit the Site 41 resistors as criminals or as troublemakers.

3.6.9   COMMUNITY RESISTANCE

Engineers described resistance to Site 41 as vocal, intense, and emotional [57]. Jagger Hims engineers pointed out that even the CMC had been taken over by passionate opponents. They described resistance to Site 41 not in terms of concerns that were brought up, but in terms of emotions. Furthermore, the engineers believed emotions had "taken over," implying that Site 41 opponents were irrational. This use of language in describing community resistance set the concerns of the community apart from the (supposedly unbiased) language of engineering and regulation used by experts.

Resistance was also talked about in terms of the political or personal. Simcoe County's landfill designers used language distancing opposition to Site 41 from the technical debate [99]. They used the language of expertise to talk about the science behind Site 41, implying that resistance was motivated by political or personal reasons [91]. For them, the facts presented by experts were not in dispute [71].

However, not all experts were equal. In describing Bill Shotyk's results, Tony Guergis, the Warden of Simcoe County, emphasized that the scientist was a local [54]. Again, resistance to the site was described as local to equate it with the political. Similarly, resistance to Site 41 was explained as a "not in my backyard" (NIMBY) issue since only local community members were opposed to the site. Any attempts by the Site 41 opponents to actually have their concerns addressed were dismissed as NIMBY opposition by Simcoe County [93]. By excusing away local concerns as NIMBY, Site 41 designers did not have to deal with criticism of their work. Formulation of community resistance as just a local problem made the concerns of local residents invisible in Site 41 design discourse.

3.6.10   INDIGENOUS PEOPLES

Simply put, Indigenous peoples were completely invisible in Site 41 design literature. They were not consulted. Their land rights were not recognized and neither was their knowledge even though the landfill site would have impacts on their treaty territories, hunting/fishing rights, and water sources. They were a subject never brought up or discussed except by Site 41 opponents themselves. The consultation responsibilities of designers at Site 41 were limited to the requirements of the Canadian Environmental Assessment Act of 1992. Since Indigenous peoples were largely absent from that legislative document,[14] the engineers could still claim to have met all their environmental and social responsibilities. By using the language of compliance, engineers cut out a crucial community from Site 41 design decisions.

---

[14]Except for a clause that considered changes made by a project "on the current use of lands and resources for traditional purposes by aboriginal persons" as an environmental effect [1].

3.7   IDENTIFYING DOMAINS, INSTITUTIONS, FORMS OF SPECIFICATION, AND POWER AT SITE 41

In the previous section, subjects, objects, and strategies in Site 41 design discourse were identified as well as the rules of formation around subjects and their relationships to each other. Engineering design discourse was shown to exclude meaningful public input, ignored Indigenous peoples, treated the environment as a passive agent, failed to acknowledge uncertainty, limited the scope of social impact analysis, allied with institutions of authority, and devalued non-expert knowledge. Two goals from the archaeological analysis of Site 41 still remain: to chart the domains, institutions, technologies, and forms of specification around the sayable and to link those domains, institutions, etc., to power relationships at Site 41.

3.7.1   DOMAINS

Domains or surfaces of emergence are the places in discourse where subjects are defined [39]. For Site 41 design discourse, the subjects were defined in the domains of waste management, landfill engineering, environmental regulation, and government legislation.

Within the domain of waste management, landfills are a necessary reality, since waste always exists. It needs to exist for waste management to exist. Furthermore, this waste needs to be managed in certain ways (diversion, containment, incineration, etc.), so landfill engineering comes into play. The domain of landfill engineering uses technical jargon and expert knowledge to create a particular discourse around constructing landfills. Meanwhile, environmental regulation and government legislation defined the scope and depth of consultation, impact assessment, and environmental protection at Site 41.

3.7.2   INSTITUTIONS

Institutions are the places in discourse that provide limits as to how a subject may exist or be meaningful [39]. At Site 41, the domains of waste management, engineering, and regulation/legislation were limited and defined by the MOE, the Canadian government, the Joint Board, Simcoe County government, engineering consulting firms, and, at times, the NSWMA. These institutions dictated the rules of consultation, the environmental standards, and engineering priorities. They also decided the evidence that was meaningful enough to be included in the discussion of the site.

3.7.3   FORMS OF SPECIFICATION

A form of specification is the way in which subjects within a particular domain are related and constructed [39]. Technical engineering jargon, management vocabulary, and bureaucratic requirements constructed subjects in Site 41 design discourse. Whether it was discussing hydraulic gradients or community consultation, everyone involved in the Site 41 design process had to use that particular language. They never considered using the language that the Anishinabe Kweag women used in describing the importance of unpolluted water and land. The institutions mentioned above played a key role in reinforcing the use of bureaucratic language and engineering jargon. In this way, the MOE and Simcoe County, along with the other institutions, were able to define how subjects at Site 41 would be talked about, not the community, farmers, or Indigenous peoples.

3.7.4   POWER RELATIONSHIPS

In looking over my analysis of Site 41 design discourse, I see that the domains, subjects, strategies, objects, institutions and forms of specification that defined the discourse were constructed to give power to engineers, experts, and state-affiliated political bodies. For example, informal/local/traditional/plant knowledge was missing as a domain in Site 41 design because engineers from the consulting firms and Simcoe County defined the domains, not local farmers or Indigenous communities. Similarly, institutions such as the CMC and other citizen's groups did not play a significant role in defining the discourse at Site 41 although they did challenge it at times. This meant that government institutions such as the MOE or Ontario Legislature got to define the rules of formation for Site 41 discourse. Even the manner in which subjects were constructed in discourse, using engineering jargon, points to the power relationship between the environment and humans. In engineering jargon, the environment was reduced from a living stakeholder to a set of inputs/outputs in a model.

3.8   TOWARDS A MORE JUST RELATIONSHIP TO WASTE IN SIMCOE COUNTY

In Chapter 2, I outlined a design process, shown in Figure 3.5, that engineers and other decision-makers could use to approach waste management in a just way. In the following sections, I engage in the exercise of applying this model to decision-making at Site 41.

3.8.1   IDENTIFY STRUCTURES OF OPPRESSION

Going back to Young's work [130] and using my analysis of power from this chapter, I can identify at least four different structures of oppression in Site 41 design: powerlessness, cultural imperialism, environmental exploitation, and violence. Citizens' groups and community members were powerless in the way debate around Site 41 was framed. Since they were not experts, they did not have access to design decisions in the same way experts such as engineers did. Indigenous communities were another stakeholder whose participation in the debate was limited, but due to cultural imperialism. Even though Site 41 impacted the land and water of Indigenous communities, the government never felt any legal or moral obligation to consult or collaborate with them. This form of oppression, according to Young, happens "when the dominant meanings of a society render the particular perspective of one's own group invisible at the same as they stereotype one's group and mark it out as the Other" (p. 58-59) [130]. The dominant perspective of Canadian settler society rendered Indigenous

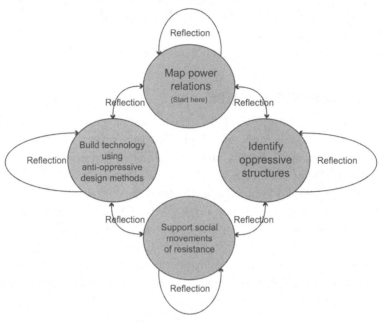

Figure 3.5: Proposed socially just design method.

communities invisible.[15] They played no part in the design decisions of Site 41. When Indigenous peoples did try to play a part in the design process through the Site 41 protest camp, they were subjected to another form of oppression, violence. Violence was deployed by the police in enforcing design decision at Site 41 through arrests, tearing down tents, and summons. Finally, the environment was constructed as something to be protected for future exploitation whether for farming or water use. The environment in Site 41 design discourse did not have any value beyond how it could be used for profit.

I should stress that each of the forms of oppression I have mentioned are fluid, cross-cutting, and shifting. While Indigenous communities were made invisible, they were also economically exploited through the construction of a landfill on their hunting and fishing lands. The environment, at times, was also made invisible through the language of regulation/compliance while, at other times, was made visible only for exploitation.

---

[15]Indigenous protesters at the Site 41 blockade were marked as "others" in a Simcoe County press release, which claimed, erroneously, that a dozen militant individuals wearing black bandanas over their faces had denied staff entry to the landfill [4]. No militants or individuals wearing black bandanas had been present at the blockade during the time indicated by the press release.

3.8.2   WORKING WITHIN SOCIAL MOVEMENTS

The WYE Citizens group, the CMC, Anishinabe Kweag, farmers (Simcoe County Federation of Agriculture [3]), Stop Dump Site 41 team, national citizens' groups (Council of Canadians), labor unions (CUPE) and Indigenous communities were just some of the groups resisting powerlessness, cultural imperialism, violence, and environmental exploitation at Site 41.[16] Engineers and other Site 41 designers could work with these communities to restructure the discourse around Site 41 to encourage participation and stop environmental/social exploitation. They could do this using tools from participatory design such as citizen panels and peer learning committees.

In particular, engineers and other decision-makers at Site 41 could have worked with the Anishinabe Kweag, the Beausoleil First Nation, the Kawartha Nishnawbe First Nation, and other Indigenous groups/communities by both acknowledging their land rights and by establishing mean-ingful participation, consultation, and veto procedures for the various Indigenous communities. Along with farmers in North Simcoe County, Indigenous communities were perhaps going to be most affected by the landfill on their lands. Not only would the placement of a landfill on their land without consent or consultation reduce their sovereignty, but it would also negatively affect their land and water-based needs (i.e., fishing, hunting, potable water, etc.).

3.8.3   USING RESISTIVE DESIGN TECHNIQUES

How could engineers have worked with these communities to resist oppression though? I propose that Site 41 designers could have drawn from four design techniques: participatory design [113] to resist powerlessness, ecofeminism [41] and deep ecology [79] to resist environmental exploita-tion/violence, and decolonized design [117] to resist cultural imperialism.

Participatory design techniques aim to meaningfully involve non-experts, community mem-bers, and users in the design process. At Site 41, participatory design could have been used to arrive at a different answer from the experts to the waste "crisis." Even without that, a more non-hierarchical approach to design decisions could have allowed non-experts to have meaningful agency in the design of the landfill. Instead of information sessions, consultations could have taken place through workshops or theater between experts/non-experts. Norms of deliberation could have been changed to make them more inclusive (moving from parliamentary procedure to sharing circles). Consultation could have started earlier in the Site 41 process through the formation of extended peer communities [40] consisting of engineers, farmers, Indigenous knowledge holders, and other local citizens.

Meanwhile, applying ecofeminist/deep ecology design principles to Site 41 could have restruc-tured the discussion around the environment from mitigation/risks/economic impact to ecosystem well-being, connectedness between people and nature, and local/plant knowledge. The knowledge of local farmers could have been made part of the hydrogeological studies conducted by the engi-neering firms. Indigenous knowledge holders could have been consulted during the environmental impact assessment. This ecofeminist-inspired design approach would have valued and encouraged

---

[16]For a complete list, check out the Stop Dump Site 41 website [9].

local knowledges and biological diversity. At the same time, design decisions at Site 41 could have been informed by the deep ecological approach to nature that "goes beyond the so-called factual scientific level to the level of self and Earth wisdom" (p. 454) [33]. Instead of viewing the environment as something to be protected under narrow guidelines, engineers could have seen it as a multitude of equal stakeholders that need to be consulted. Independent advocates for the environment, much like the Environment Commissioner of Ontario but on a more local scale, could have represented the environment in design decisions. Organizations with specific experience (fishing preservation or local plant species) could sit on a peer panel together with local citizens to write environmental impact reports. This would have ensured that the environment was adequately recognized and represented as an active stakeholder from various communities.

Finally, the design process at Site 41 needed to be decolonized so that Indigenous communities, knowledges, and rights could be recognized. The design process at Site 41 privileged scientific ways of knowing and material progress over nature or the relationship between people and their land. In contrast, an Indigenous nature-based design would value "reciprocity or reciprocal relations that define responsibilities and ways of relating between humans and the world around them" (p. 202) [110]. It makes sense then that such a design would be "responsive and responsible to the ecology in which it lives and from which it came" (p. 6) [125]. This Indigenous-based design could have been created through collaboration with Indigenous communities or through partnerships with Indigenous elders and other knowledge holders. In addition, a decolonized design at Site 41 would have recognized the hunting/agricultural lands, aquatic sources, heritage sites, and knowledges of Indigenous communities affected by the landfill.

3.9   TOWARDS ZERO WASTE IN SIMCOE COUNTY

Towards the end of the Site 41 debate in 2009, Simcoe County hired an engineering consulting firm, Stantec Consultants Inc., to develop a waste management strategy for the county that included a path towards zero waste [96]. This happened largely because of the pressure from opposition to Site 41. Non-experts, farmers, and Indigenous communities all banded together to create a new vision for how their region handles waste by rejecting the Site 41 landfill and participating in the creation of a waste management strategy. Stephen Ogden pointed out the irony of the situation though by noting how the WYE Citizens group had submitted a 10 point plan to the second Joint Board hearings on Site 41 [104]. The plan had called for zero waste among other goals. It had been rejected. Almost 15 years later, many of the same points from the WYE submission were made in the final Simcoe County Waste Management Strategy released in 2010 [97].

REFERENCES

[1] Canadian environmental assessment act, 1992. 53

[2] Canadian issues in environmental ethics, chapter Terrorism at Oka. Broadview Press, 1997. 30

[3] Meeting of committee of the whole, 2007. 57

[4] County news release racist, native protester says, September 3, 2009. 56

[5] Part a simcoe county info session dump site 41 August 25, 2009, 2009. 53

[6] Part b simcoe county info session dump site 41 August 25, 2009, 2009.

[7] Stop Dump Site 41. Media: Documents, videos, and photos, 2009. `http://stopdumpsite41.ca/?page_id=8` 39

[8] From Stop Dump Site 41. Huge rally attracts 2,500, 2009. 41

[9] From Stop Dump Site 41. Supportive groups and individuals, 2009. `http://stopdumpsite41.ca/?page_id=10` 57

[10] Stop Dump Site 41. A brief history and timeline of events, 2010. `http://stopdumpsite41.ca/?page_id=2` 33, 35

[11] Juli Abouchar. Ipc order mo-2449/appeal ma07–365 - institution file number: A17-5520-07 - county of simcoe - your file no. g15553, 2009. 41

[12] Anonymous. Metis nation of ontario opposes site 41, June 22, 2009. 38

[13] Anonymous. Site 41 arrests point to "disturbing" trend of criminalizing legitimate protest, warns cupe ontario president, August 10, 2009. 39

[14] John Bacher. Mohawk activist leads walk for water, January 2009. 38

[15] Jason T. Balsdon and Andrew G. Hims. Re: County of simcoe: Landfill site 41 - reply to county letter dated June 15, 2009 (file 880007.28), 2009. 39, 50

[16] Frank S. Barone. Peer review of final design for county of simcoe landfill - site 41 (tiny township). Technical Report 03–1113-004, Golder Associates Ltd., Urban and Environmental Management Inc., Dr. R.K. Rowe Inc., 2003. 36, 48

[17] Lalita Bharadwaj, Suzie Nilson, Ian Judd-Henrey, Gene Ouellette, et al. Waste disposal in first-nations communities: The issues and steps toward the future. Journal of Environmental Health, 2006. 31

[18] Colin Bhattacharjee. Order mo-2416/appeal ma07–365. Technical report, Information and Privacy Commissioner/Ontario, 2009. 39

[19] Colin Bhattacharjee. Order mo-2449/appeal ma07–365. Technical report, Information and Privacy Commissioner/Ontario, 2009. 40

[20] Colin Bhattacharjee. Re: Stay decision - orders mo-2416 and mo-2449 - appeal ma07–365 - institution file number: A17-5520-07, 2010. 41

[21] Raymond Bowe. Pauze dump still raising concerns 22 years later, 2010. 32

[22] Linda Bruce and Kate Harries. The Anishinabe Kweag were bound to protect the water for future generations, 2010. 38

[23] S. Carter. Aboriginal people and colonizers of Western Canada to 1900. Themes in Canadian social history. University of Toronto Press, 1999. 29

[24] Jill Colvin. Opp takes down tents at landfill protest, August 8, 2009. 39

[25] Community Monitoring Committee. Minutes of meeting #12–05 held December 8. Technical report, 2005. 51

[26] Community Monitoring Committee. Landfill site #41 - 2008 annual report. Technical report, CMC, 2008.

[27] Community Monitoring Committee. Minutes of meeting #02–08 held February 14. Technical report, 2008. 37, 47

[28] Community Monitoring Committee. Minutes of meeting #3–08 held March 13. Technical report, 2008. 37

[29] Community Monitoring Committee. Minutes of meeting #4–08 held April 10. Technical report, 2008. 51

[30] Ontario Executive Council. Re: Application to the lieutenant governor in council (the "lgic") by the north simcoe waste management association ("nswma") by way of appeal from a decision of the joint board re tiny township landfill site, June 14, 1990 1990. 34

[31] Adrian Coyle. Discourse Analysis, pp. 98–116. Analysing Qualitative Data in Psychology. SAGE Publications, Ltd, 2007. 42

[32] Guy Crittenden. A site to behold, March 1 2005. 47, 49, 51

[33] Bill Devall and George Sessions. Deep Ecology, chapter 38, pp. 454–467. Technology and Values: Essential Readings. Wiley-Blackwell, 2010. 2008054060. 58

[34] Paul Dewaele, Rick Mosher, and Keith Sherman. County of simcoe landfill site 41: Peer review of hydrogeology and landfill design reports. Cs 03–157, Dixon Hydrogeology, Conestoga Rovers and Associates Ltd., Severn Sound Environmental Association, 2003. 35, 36, 48

[35] Christina Dhillon and Michael G. Young. Environmental racism and first nations: A call for socially just public policy development. Canadian Journal of Humanities and Social Sciences, 2010. 30

[36] John Edwards. Council votes to revoke c of a for site 41, May 25, 2010. 42

[37] Robert B. Eisen. Re county of Simcoe landfill site, 1996. 34, 47, 49, 50, 51, 52

[38] Robert B. Eisen and Dorothy H. McRobb. Reasons for decision and decision, 1989. 32, 33, 48, 49, 51, 52

[39] M. Foucault. Archaeology of knowledge. Routledge, 2007. 2002067999. 43, 44, 45, 54

[40] Silvio O. Funtowicz and Jerome R. Ravetz. Science for the post-normal age. Futures, 25(7):739, 1993. DOI: 10.1016/0016-3287(93)90022-L 57

[41] G. C. Gaard. Ecofeminism: women, animals, nature. Temple University Press, 1993. 92006598. 57

[42] Tesfaye Gebrezghi. Revocation of provisional certificate of approval waste disposal site, 2010. 42

[43] Douglas Glynn. Site 41 battleground expands; queen's park new front in war against dump, 2008. 37

[44] Douglas Glynn. Walk away: First nations artist Danny Beaton is walking in opposition to site 41. he says when it comes to the dump site, Simcoe County should, November 12, 2008. 38

[45] Douglas Glynn. Cmc a step closer to key site 41 data, 2009. 37

[46] Douglas Glynn. Cops raid site 41, 2009. 39

[47] Douglas Glynn. Residents walk for their water, May 5, 2009. 38

[48] Douglas Glynn. Site 41 protesters may face charges, 2009. 39

[49] Douglas Glynn. Charges stayed against site 41 protesters, 2010. 39

[50] Douglas Glynn. County's legal bills in modflow dispute said nearly 60,000, April 15 2010. 37

[51] Kim Goggins. Work underway at site 41, August 30 Kim Goggins-Aug 30, 2007. 37

[52] E. Marshall Green. Re: Continuation of inquiry - order mo-2416/appeal ma07–365 - our file g15553, 2009. 40

[53] E. Marshall Green. Re: Ipc order mo-2449/appeal ma07–365 - institution file number: A17-5520-07 - county of simcoe - our file g15553, 2009. 40

[54] Tony Guergis. Public being misled concerning dump at site 41, 2010. 47, 50, 53

62 REFERENCES

[55] Kate Harries. Interview with William Shotyk, 2009. 36

[56] D. E. Jagger. Re: County of simcoe landfill site 41 - hydrogeological and geotechnical aspects - response to ministry of the environment comments - file 880007.11, 2006. 48, 51

[57] D. E. Jagger and Jason T. Balsdon. Twenty five years and counting: public perceptions, attitudes, and actions. In Waste—The Social Context, pages 282–287, May 11–14, 2005 2005. 47, 48, 49, 50, 51, 52, 53

[58] B.E. Johansen and B. Pritzker. Encyclopedia of American Indian history. Number v. 1 in Encyclopedia of American Indian History. ABC-CLIO, 2008. 31

[59] Marianne Jorgensen and Louise J. Phillips. Discourse Analysis as Theory and Method. SAGE Publications Ltd., 2002. DOI: 10.4135/9781849208871 42

[60] James D. Kingham. County of Simcoe landfill site decision and reasons for decision, 1992. 34

[61] Glen R. Knox. Chronology: Request for information under the municipal under the municipal freedom of information and protection of privacy act ("calibrated usgs modflow model"). Technical Report CO 09–027, 2009. 37

[62] Glen R. Knox. Re: County of Simcoe: Landfill site 41, 2009. 39

[63] Glen R. Knox and Rob McCullough. Ministry of the environment sponsored meetings – site 41 modflow model. Technical report, 2008. 37

[64] Henderson Paddon and Associates Ltd. Development and operation report – Simcoe landfill site no. 41, tiny township. Submission to moe, 2003. 35

[65] Henderson Paddon and Associates Ltd. Landfill base preparation test trial – summary report - county of Simcoe landfill site no. 41, tiny township. Prepared for County of Simcoe Environmental Services Department 199719, 2007. 52, 53

[66] Terraqua Investigations Ltd. Assessment of closed e. pauze landfill site. Technical report, Queen's Printer for Ontario, 1992. 32

[67] Rob McCullough. County landfill site 41 information publication. Technical Report CO 04–014, 2004. 49

[68] Rob McCullough. Site 41 – county infrastructure requirement. Technical Report CS 06–271, 2006. 47, 52

[69] Rob McCullough. Site 41 cmc motion 12/13/07 #1 re: request for funding to peer review the december 2007 report on landfill base preparation test trial report. Technical Report CS 08–021, 2008. 49

[70] Rob McCullough. Site 41 cmc motion 12/13/07 #2 re: request for funding to peer review the county's application for a permit to take water. Technical Report CS 08–022, Simcoe County, 2008. 37

[71] Rob McCullough. Site 41 – notice of motion. Technical Report CO 06–048, 2009. 53

[72] Christie McLaren. Ratepayers irked by secrecy of tiny landfill site committee, August 25, 1984. 32, 33

[73] David Merriman. Business case analysis for county landfill site 41. Prepared for County of Simcoe MA-07–006-00-MA, Genivar Ontario Inc., 2007. 50

[74] Ray Millar. Re: Application for permit to take water/ lots 10 and 11 conc. 2 twp. of tiny (site 41), 2008. 37

[75] Ray Millar. Re: Proposed county of Simcoe landfill site 41, 2008. 36, 37

[76] Nicole Million. Town baffled by presence of toxic chemicals in well, May 11, 2010 2010. 32

[77] Martin Mittelstaedt. Earth's cleanest water creates thorny issue, December 16, 2006. 36

[78] Martin Mittelstaedt. Protestors ordered to end blockade at landfill construction site, July 23, 2009. 38

[79] Arne Naess. The shallow and the deep, long-rage ecology movement. A summary. Inquiry, 16(1–4):95–100, 1973. 57

[80] Kris Nahrgang. Re: Dump site 41, 2009. 38

[81] John J. Nahuis. Regarding: County of Simcoe's (site 41) pttw-permit to take water, 2008. 37

[82] Council of Canadians. Coalition launches facebook campaign to stop dump site 41; demands mcguinty government reverse decision to allow dump site on pristine aquifer, 2009. 38

[83] Council of Canadians. Council of canadians demands halt to construction of waste dump, 2009. 38

[84] Council of Canadians. New conflict of interest confirms site 41 is out of control, says council of canadians, July 14, 2009. 40

[85] Council of Canadians. Site 41 debate heads from the courtroom back to council chambers, 2009. 39

[86] Council of Canadians. Stop site 41, 2009. 40, 41

[87] Council of Canadians. Stop site 41: No water to waste, 2009. 38

64 REFERENCES

[88] Council of Canadians. Update: Simcoe County council vote on site 41, 2009. 41

[89] Council of Canadians. Win! Simcoe County council votes "to cease construction and all further development of site 41," 2009. 42

[90] Environmental Commissioner of Ontario. Annual report supplement: Developing sustainability. Technical report, 2001/2002. 35

[91] Environmental Commissioner of Ontario. Annual report: Choosing our legacy. Technical report, 2003/2004. 36, 53

[92] Environmental Commissioner of Ontario. Home, 2010. 35

[93] County of Simcoe. Site 41 update: Waste management strategies in the county of Simcoe. Technical report, 2005. 47, 51, 53

[94] County of Simcoe. Landfill site 41 receives final ministry design plan approval. Press release, 2006.

[95] County of Simcoe. Managing your waste. Technical report, 2006. 47, 50

[96] County of Simcoe. County of Simcoe new solid waste management strategy. Press release, 2009. 58

[97] County of Simcoe and Stantec Consulting Inc. Solid waste management strategy. Technical report, 2010. 58

[98] Environmental Service (County of Simcoe). County of Simcoe waste management workshop, 2007. 47

[99] Environmental Services (County of Simcoe). Why landfill site 41 is required. Technical report, Simcoe County, 2006. 47, 53

[100] Environmental Services (County of Simcoe). Managing your waste fact sheet, 2009. 48, 50, 52

[101] Office of the Auditor General. Sanitary landfill site development proposal in Simcoe County, Ontario, 2005. 37

[102] Ministry of the Environment. Certificate of approval, 2000. 35

[103] Ontario Ministry of the Environment. North Simcoe waste management association landfill environmental assessment joint board decision: An issues analysis. Part of a ministry funded project, 1990. 33, 49, 50, 52

[104] Stephen Ogden. The why wye group's vision 20 years ago – can we run with it now?, 2010. 35, 58

[105] Ogilvie and Company Ogilvie. Site 41 modflow facilitator's summary, 2008. 48, 49, 50

[106] James O'Mara. Re: The ministry of the environment has received an application for a review of the director's decision to issue certificate of approval no. a602278 for Simcoe landfill site 41, in the township of tiny. EBR Application for Review 04EBR001.R, 2004. 35, 36

[107] Brent Patterson. Council criticizes death of site 41 bill due to prorogation of ontario legislature, 2009. 42

[108] Kimberley Pickett and Rob McCullough. North Simcoe landfill site (site 41) update. Technical Report CO 09–025, Simcoe County, 2009. 37

[109] Jim Rankin. Where worlds collide: The water gushing from the land called site 41 has been described as perhaps the cleanest on earth so why does the county of Simcoe want to cover this resource with a landfill?, April 6, 2006. 36

[110] Linda Robyn. Indigenous knowledge and technology: Creating environmental justice in the twenty-first century. The American Indian Quarterly, (2):198, 2003. DOI: 10.1353/aiq.2003.0028 58

[111] R. Kerry Rowe. Site 41 modflow modelling, 2008. 51

[112] LM Schell, LA Hubicki, AP DeCaprio, MV Gallo, J Ravenscroft, A Tarbell, A Jacobs, D David, P Worswick, and Akwesasne Task Force on the Environment. Organochlorines, lead, and mercury in akwesasne mohawk youth. Environmental Health Perspectives, 2003. DOI: 10.1289/ehp.5990 31

[113] Richard Sclove. Democracy and technology. The Guilford Press, New York, US, 1995. 57

[114] William Shotyk and Michael Krachler. Determination of trace element concentrations in natural freshwaters: How low is "low", and how low do we need to go? Journal of Environmental Monitoring, 11:1747–1753, 2009. DOI: 10.1039/b917090c 36

[115] William Shotyk, Michael Krachler, Werner Aeschbach-Hertig, Stephen Hillierc, and Jiancheng (James) Zheng. Trace elements in recent groundwater of an artesian flow system and comparison with snow: enrichments, depletions, and chemical evolution of the water. Journal of Environmental Monitoring, 12:208–217, 2010. DOI: 10.1039/b909723f 36

[116] William Shotyk, Michael Krachler, Bin Chenz, and James Zhengy. Natural abundance of sb and sc in pristine groundwaters, springwater township, ontario, canada, and implications for tracing contamination from landfill leachates. Journal Environmental Monitoring, 7:1238–1244, 2005. DOI: 10.1039/b509352j 36

[117] L. T. Smith. Decolonizing methodologies: research and indigenous peoples. Zed Books, 2006. 57

[118] Dave Staseff. Supplemental hydrogeological and geotechnical investigation landfill site 41 - county of Simcoe - conditions 9, 10.1(c), and 10.1(d) - certificate of approval (c of a) no. a620278. Technical report, Ministry of the Environment, 2003. 35, 51

[119] Gail Swainson. Activists hope to beat the odds in dump dispute, August 24, 2009. 38

[120] Peter Tabuns. Waste disposal, 2009. 47

[121] Peter Tabuns. Waste disposal, 2010. 48, 50

[122] Environmental Review Tribunal. About the office of consolidated hearings, 2011. 33

[123] Teun A. van Dijk. Principles of critical discourse analysis. Discourse and Society, 4(2):249–283, 1993. http://das.sagepub.com/content/4/2/249.full.pdf+html DOI: 10.1177/0957926593004002006 43

[124] G.R. Vizenor. Landfill meditation: crossblood stories. Wesleyan University Press, 1991. 29

[125] Rose von Thater-Braan. Is it possible to have information technology that reflects indigenous consciousness? Technical report, 2007. 58

[126] Sharon Weatherall. Protestors fighting dump proposed over pristine water, May 2009. 47, 48

[127] Ruth Wodak. What CDA is about – a summary of its history, important concepts and its developments, chapter 1, pages 1–13. Methods of Critical Discourse Analysis. SAGE Publications, Ltd., 2001. 43

[128] Ruth Wodak. Critical Discourse Analysis, chapter 12, pages 186–198. Qualitative Research Practice. SAGE Publications Ltd, 2004. 43

[129] Robin Wooffitt. Conversation Analysis and Discourse Analysis. SAGE Publications, Ltd, 2005. 42, 43

[130] Iris Marion Young. Justice and the politics of difference. Princeton University Press, Princeton, New Jersey, US, 1990. 55

[131] Duane Champagne. Notes from the center of Turtle Island. AltaMira Press, Lanham, Maryland, US, 2010, p.45–46. 29

[132] Indigenous Environmental Network. http://www.ienearth.org/. Accessed December 2012. 29

[133] Christina Dhillon and Michael G. Young. Environmental Racism and First Nations: A Call for Socially Just Public Policy Development. Canadian Journal of Humanities and Social Sciences, vol. 1, is 1: 25–39, 2010. 29

[134] Beverley Jacobs. Environmental racism on Indigenous lands and territories. Canadian Political Science Association Annual Conference, 2010. 29

[135] Laura Westra. Environmental Justice and the Rights of Indigenous Peoples. London, UK: Earthscan, 2008. 29

[136] Elaine MacDonald and Sarah Rang. Exposing Canada's Chemical Valley: an investigation of cumulative air pollution emissions in the Sarnia, Ontario area. Ecojustice: October 2007. 30

[137] C. A. Mackenzie, A. Lockridge, and M. Keith. Declining Sex Ratio in a First Nation Community. Environmental Health Perspectives, 113 (10): 1295–8. 2005. DOI: 10.1289/ehp.8479 30

[138] Environmental Law Centre at University of Victoria. Environmental Rights: human rights and pollution in Sarnia's Chemical Valley. June 2011. Accessed in December 2012: http://www.elc.uvic.ca/associates/documents/ChemicalValleyAssociatesBackgrounder_June13.11.pdf 30

[139] Christopher Vecsey. Grassy Narrows Reserve: Mercury Pollution, Social Disruption, and Natural Resources: A Question of Autonomy. American Indian Quarterly, vol. 11, no. 4, 1987, pp. 287–314. DOI: 10.2307/1184289 30

[140] April Kinghorna, Patricia Solomona, Hing Man Chanb. Temporal and spatial trends of mercury in fish collected in the English–Wabigoon river system in Ontario, Canada. Science of The Total Environment, vol 372, issues 2–3, January 2007, pp. 615–623. DOI: 10.1016/j.scitotenv.2006.10.049 30

[141] Harada, M., et al. Mercury Pollution in First Nations Groups in Ontario, Canada: 35 years of Canadian Minamata Disease. Journal of Minamata Studies 3, 2011, pp. 3–30. 30

[142] Free Grassy Narrows. http://freegrassy.org/. Accessed in December 2012. 30

[143] Maura Hanrahan. Brooks, Buckets, and Komatiks: The Problem of Water Access in Black Tickle. Division of Community Health, Memorial University of Newfoundland, 2000, p. 2. 30

[144] Chief Rodney Monague Jr. Letter to Simcoe County from Beausoleil First Nation. August 11, 2009. stopdumpsite41.ca. 38

[145] Icarus Project. http://theicarusproject.net/. Accessed on December 24, 2012. 43

CHAPTER 4

Waste Management in the Global North

Randika Jayasinghe

Wastes often follow the channels of the global economy as vehicles to disadvantaged communities, taking advantage of economic inequality to "solve" the waste problem. (p. 173)

Jennifer Clapp [1]

4.1 INTRODUCTION

Waste management is a major environmental burden throughout the world, including in economically developed countries. One can argue that a "garbage crisis" does not exist in developed countries, as most would expect the waste to be collected and taken away from their houses. However, the increasing amounts of waste generated each year in developed countries have raised concerns about the overall sustainability of the current waste disposal methods in the global North [2]. Developed countries are generating more and more waste at an alarming rate spurred by urbanization, industrialization, and modernization. Currently, world cities generate about 1.3 billion tonnes of solid waste per year. This volume is expected to increase to 2.2 billion tonnes by 2025 [3].[1] Developed countries contribute largely to this with waste quantities generated by developed countries been increased at a rate of 3% per year over the last few decades [4].

Waste is certainly a problem in the global North although there are not many noticeable evidences to support this statement. Most people tend to ignore the fact that there is a "waste problem" in developed countries, partly because the different features of development such as highways, skyscrapers and large industrial developments have concealed the issue. Masked by the beauty of development, problems related to managing waste in developed countries often affect low-income communities, including low-income countries around the world.

The waste management processes in most developed countries are modernized to serve the demands of the population [5]. According to Scheinberg (2008), modernization of waste management represents a system where waste is removed from populated areas and dumped in some vacant or far-away place [5], either in the same country or in another. This remote destination will then create another chain of processes for waste; sorting, separating and directing the waste to appropriate

[1]The global averages are broad estimates as rates vary considerably by region, country, city, and even within cities.

intermediate or final destinations. This chapter will raise and address questions on how and why waste generated in high-income countries affects communities in these "remote" places. We are interested in understanding and sharing our insights by analysing various waste issues using a social and environmental justice lens.

Looking through a social and environmental justice lens allows to raise different questions about waste management compared with conventional discussions. In this chapter, we will focus on a new way of looking at waste management in the global North. Waste management in developed countries seems perfect to those from developing countries. Yet, as demonstrated in the last chapter, evidence suggests that part of the problem is shifted to low-income communities or countries, especially when it is easy and economical to pass the responsibility to someone else [6, 7, 8].

To address some of these unseen dilemmas of waste management in developed countries, this chapter attempts to answer the following questions.

- What are the key elements of a modernized waste system?

- What governs the waste management system of developed countries?

- What happens in the solid waste process in developed countries, and what drives it?

- How has this affected different stakeholders in waste management?

- How is hazardous waste managed?

- How and why are low-income communities and countries affected?

High-income countries and cities are usually seen as environmentally conscious places. Residents of these countries usually separate garbage into different categories according to their recyclability. There are designated places to drop off electronic waste (e-waste) and to collect hazardous waste. One can come to believe that this is a perfect system. Indeed, it is a perfect system if we look through a consumption lens; the cities look clean and garbage is taken care of by the city councils or private companies.

It is important to note what happens to all the waste once it is taken away. A portion of it is buried in landfills, incinerated, or converted to compost. Some waste is recycled within the country and a considerable amount is shipped to developing countries for further processing [9]. Exported waste includes hazardous and e-waste that is not landfilled or treated due to the strict environmental regulations in developed countries [10]. Our emphasis will be on this global trade of "dumping on others," particularly on marginalized groups such as low-income communities and ethnic minorities in both developing and developed countries.

We used the lens of social and environmental justice to critically analyze some of the diverse and complex waste management practices of the global North and we encourage our readers to do the same. If we take a step back and look at waste management in developed countries, we can raise a critical question: Can we consider the waste management in developed countries a perfect one, when "dumping on others" is taken as a legitimate method of waste management? We hope that this chapter will encourage our readers to think beyond what we already know. We conclude this chapter with the thought that solving the waste problem means more than passing waste off to people in other places.

4.2   WASTE MANAGEMENT IN THE GLOBAL NORTH

"What appear to be improvements in environmental quality may in reality be indicators of increased ability of consumers in wealthy nations to distance themselves from the environmental degradation associated with their consumption."(p. 177)

Dale S. Rothman [11]

Globalization and industrialization have created enormous barriers between economic development and human well-being. The limited resources on earth are being depleted at an alarming rate while pollution and related environmental issues have increased dramatically. There is a clear separation between human activities and the environmental impact they cause, making it much harder to trace the source of the problem. We place more emphasis on economic growth, industrial development, and creating global networks than preserving our moral values and ethics. The same attitude has led to externalizing the cost of waste.

Most residents in developed countries tend to neglect the waste problem that is choking the planet. Many would express concern about the environment yet continue to live materialistic lifestyles that result in generating more waste [12]. There is a growing waste problem in the industrially developed countries even though many of the residents do not see or experience it firsthand. The main reason behind this "do not care" attitude is that waste is considered as an essential service of the local authority. Residents do not bother about managing waste beyond their garbage bin and often claim that the tax payments should deal with the issue.

Many do not seem to understand the link between high consumption and waste generation. Residents in developed countries consume far more than they already require [12, 13]. For example, developed nations which comprise only 16% of the world's population consume approximately 75% of global paper production [13]. According to a survey carried out in 2004, overall, Australians had thrown away $2.9 billion of fresh food, $630 million of uneaten take-away food, $876 million of leftovers, $596 million of unfinished drinks, and $241 million of frozen food; a total of $5.3 billion on all forms of food in 2004 [12]. Today, after eight years, these amounts can be much higher.

When purchasing goods is undertaken without proper consideration or as a form of a hobby, wasted resources cannot be seen as by-products of what we consume [12]. Purchasing products that we really do not need in the first place inevitably leads to more waste being added to the environment, including partially used and unused items. Nevertheless, most consumers in developed countries hardly question their waste, not when it is conveniently and regularly taken away from their residential areas.

Local authorities consider waste management to be one of the most important services provided to the residents [14]. At present, private and government involvement is high in managing waste whereas informal sectors are almost non-existent. The modernization of the waste system in many developed countries led to the removal of the informal sectors from the waste management system, making way for private companies [5]. Today, waste management has become a profitable business for many private companies.

When the modernization process started in developed countries in the 1970s, waste management was seen largely as a technical problem which required technical solutions. However, the modernization of waste management in developed countries in recent decades saw several significant changes [5]. Small local landfills were closed and regional landfills were opened. The costs for collection, transfer, and transport of waste increased due to regionalization. A greatly expanded interest in recycling, composting, and recovery approaches arose for municipal service providers, both private and government owned. This led to the prohibition and/or criminalization of traditional solid waste practices, especially those relating to informal waste activities [5].

Technologies developed in the global North are designed for their own local circumstances where labour costs and technical capacities are high [4, 15]. Moreover, the composition of waste in developed countries is different to that of developing countries. Residents of developed countries consume more processed foods, resulting in more packaging materials in the waste [15]. The waste stream is also rich in discarded electronic items [16] and industrial waste. These different waste streams, which include electronics, plastics, and hazardous waste, pose special waste disposal challenges. Managing some of these waste in a socially and environmentally acceptable manner will form the basis for our discussion.

We raise one key question to initiate our discussion: What does waste management mean for a country in the global North?. We have looked at this based on the following three aspects:

- distancing of waste;

- dumping in marginalized communities within developed countries; and

- dumping in developing countries.

We highlighted different issues in relation to waste management in a developed country and one of the most important concerns is the growing distance between people and their waste. In this chapter, we draw the attention to this distancing of waste in developed countries. The distance is growing physically and mentally between people and the waste they produce [1, 17]. According to Clapp (2002), as waste distancing increases, people have little understanding of where their waste ends up, increasing the tendency to add more waste to the environment [1]. When people are not well aware about the social and ecological impacts of their waste, there is less motivation to change their attitudes.

Economic globalization has created paths for wastes to be transported away from many developed countries to developing countries [18]. This has led to developed countries feeling less responsible for the waste they create, often neglecting the social and environmental values of people in developing countries. This situation can also be seen within the same country where the waste is generated [7]. Although waste management is seemingly perfect in the metropolitan areas of many developed countries, the situation can be much different in rural and remote locations, and in areas where low-income and minor ethnic groups are residing [19]. Pellow (2004) states that numerous studies of the intersection between environmental hazards and community demographics conclude that environmental inequality is prevalent in communities across the U.S. [20]. A similar study carried out by Gawande et al. (2000) on the internal migration of U.S. hazardous waste sites provide empirical evidence that wealthy households and individuals are distancing themselves from pollution [21]. Similarly, new waste sites cannot afford to be situated near wealthy neighborhoods as gentrification increases land values such that the waste sites have to move out elsewhere; in most instances to less expensive areas where poor people live.

Rothman (1998) argues that solving environmental problems associated with growth must mean more than "passing them off" to people in other times and places [8]. Based on this, it can be argued that polluting industries and their associated waste problems should migrate away from the low-income communities and countries to wealthier ones. The latter have stricter standards which could, in effect, lead to a lowering of pollution and resource degradation levels [22]. Developed communities and countries have more capital and advanced technology to manage their waste. Hence, managing waste should not be at the cost of the lives of marginalized people in a different place.

However, in a highly consumerist society, people do not think about questions that do not affect them directly such as:

- Who is handling waste?

- Where are all these waste dumps located?

- Is anyone affected in these away places?

- What happens to all the e-waste and hazardous waste?

- Why is it that e-waste and hazardous waste is exported to developing countries, if the waste management technology is advanced in developed countries?

- What are the laws and regulations governing waste movement across borders? How effective are these regulations?

In this chapter, our key discussion will be on "cost externalization" of waste by dumping it on marginalized communities, especially with regard to hazardous waste. The relationship between globalization and transboundary movement of waste and the impact on marginalized communities both in social and environmental quality terms, is a widely discussed topic. We continued on this same discussion, using the environmental and social justice lens.

4.3   ENVIRONMENTAL INEQUALITY: SITING OF WASTE FACILITIES

"Unequal interests and power arrangements have allowed poisons of the rich to be offered as short term remedies for poverty of the poor."(p. 165)

Robert D. Bullard [19]

Environmental and social inequality occurs locally, where marginalized communities are disproportionately burdened by waste facilities and polluting industries, and globally where hazardous wastes are shipped from rich to poor countries. In this section, we will discuss the former scenario; where unequal interests play out domestically within the communities of the same country. Many research studies carried out throughout the world have concluded that low-income and vulnerable communities are paying the price of environmental hazards [6, 20]. We have drawn on case studies in two developed countries to show that environmental inequalities are prevalent in the global North, particularly when it comes to siting the waste facilities. This was explored in detail in the last chapter for Site 41 in Canada.

4.3.1   WASTE DUMPS IN THE U.S.

Environmental inequalities with regard to siting the waste facilities exist across the U.S. [6, 7, 20, 23]. Pellow (2004) argues that "communities with low levels of voting behavior, home ownership, wealth, and disposable income" and "communities of color" (p. 513) frequently have to deal with poor environmental conditions, "including polluting industries, landfills, incinerators, and illegal dumps" (p. 511) [20]. In his study "Dumping in Dixie," Bullard demonstrated that polluters prefer to choose communities who are least resistant to having a dumpsite in their neighborhoods [19]. His findings on solid waste sites in Houston revealed that most of the city's municipal landfills were located in predominantly African American neighborhoods and that white political leaders have regarded African American neighborhoods as the most appropriate sites for waste disposal [6].

In a recent study of the state of Massachusetts, Faber and Krieg (2001) analyzed both income-based and racially based decisions to the siting of 17 different types of environmentally hazardous sites and industrial facilities [23]. Their findings indicated that these hazardous sites and facilities, "ranging from highly polluting power plants to toxic waste dumps," are disproportionately located in minority working-class communities (p. iii) [23]. In a similar study in Los Angeles County, Pastor, Sadd and Hipp (2001) found that toxic waste storage and disposal facilities were disproportionately located in minority communities. Most of these facilities are businesses that accept waste from other industries. They also found that "disproportionate siting" of waste facilities occurred more than "disproportionate minority move-in" in the area [7]. Moreover, nationally, three of the five largest commercial hazardous waste landfills are found in predominantly African American and Latino communities. As a result, the residents of these communities reported suffering from poor health and an overall degraded quality of life [24].

In a case study of illegal dumping in Chicago's West side, Pellow (2004) found that more than 80% of the city's formal garbage disposal occurs in landfills in the working-class suburbs where African American, Latino, and European ethnic groups are residing. Chicago hosts the most landfills per square mile in the nation [20]. Yet, informal or illegal dumping is a common practice. One of the most publicised cases was the "Operation Silver Shovel," an illegal dumping scandal that occurred in Chicago's West Side Latino and African American communities during the 1990s [20]. Thousands of tons of debris from construction, demolition, and residential remodeling projects were dumped in these communities. The operator paid bribes to local council members in order to continue without any disruption. The bribes reinforce the observation that political leaders are willing to gain private benefits by supporting illegal dumping activities. It also highlights working-class residents' vulnerability, political powerlessness, and the social inequality prevalent in the society.

4.3.2 WASTE MANAGEMENT IN REGIONAL AND REMOTE AREAS OF AUSTRALIA

One third of Australians live in regional, remote, and Indigenous communities [25]. The remoteness of the areas generate different challenges for the relevant local authorities, which includes proper landfill management, lack of recycling facilities in the remote areas, waste management in Indigenous communities, and management of mining waste in some locations [26]. According to the National Waste report 2010, the local councils in these areas noted that the waste they manage is similar to what is generated in the metropolitan areas in terms of nature, complexity, and hazardousness [26]. This suggests that regional and remote locations require the same attention in managing their waste although the volumes can be much lower than their urban counterparts.

Most small-to-medium scale landfills are found in regional and remote areas [26]. A small landfill receive less than 10,000 tonnes per year and a medium-scale landfill has the capacity to receive up to 100,000 tonnes per year [26]. Many of these landfills may not be suitable to enable gas capture due to their size, scale, topography, and location [27]. Moreover, application of landfill liners and leachate collection systems were low in small landfills [26, 28]. Landfills with a capacity of over 100,000 tonnes are highly likely to have these features to reduce the impact on air, water and soil quality, and are mostly sited close to metropolitan locations [29].

The "Review of the Application of Landfill Standards" report states that "overall, the applications of design and construction requirements that comply with landfill guidelines have been satisfactory for large landfills, marginal for medium landfills, and unsatisfactory for small landfills" (p. 08) [28]. The number of these small landfills is relatively large (at 262 landfills accounted for in the WMAA National Landfill Survey). According to the same report, there are many unlicensed unattended tips that add to this list. Hence, when all are taken into consideration, the potential threat to environment and human health is geographically widespread, but collectively significant [28]. Local councils in remote locations may find it difficult to manage landfills and recycling centres with the limited amount of funds received from the state government and from the ratepayers. However, this may not be a valid justification not to manage waste in an environmentally and socially just manner.

Another major drawback in waste management in regional and remote locations is the lack of local recycling facilities and markets for recovered materials [26]. Kerbside collection of recyclables is available in some regional areas; however, the recycling facilities are generally clustered around major cities or regional centres. According to the South East Resource Recovery Regional Organization of Councils (SERRROC) in South-Eastern New South Wales, councils are vulnerable when it comes to choosing between the cost vs. the recycling option [30]. This suggests that the costs of collecting and transporting materials to a recycling centre in an urban area can be much higher than simply disposing in a local landfill. Similarly, the City of Darebin in Melbourne has estimated that it would cost about $1000 per tonne to recycle televisions, compared with the cost of $45 per tonne to dispose in a landfill [31] which may encourage the council to choose the more economical option. Monetary resources play a critical role in today's world; thus, there is a high incentive for local councils to implement low-cost solutions, even though those practices may cause threats to the environment and people.

4.4   HAZARDOUS WASTES: SHIFTING THE RESPONSIBILITY

"Wherever we live, we must realize that when we sweep things out of our lives and throw them away....they don't ever disappear, as we might like to believe. We must know that "away" is in fact a place. In a world where cost externalization is made all too easy by the pathways of globalization, "away" is likely to be somewhere where people are impoverished, disenfranchised, powerless and too desperate to be able to resist the poison for the realities of their poverty. "Away" is likely to be a place where people and environments will suffer for our carelessness, our ignorance, or indifference."(p. 6)

Jim Puckett [18]

Hazardous waste is defined as a "component of the waste stream which by its characteristics poses a threat or risk to public health and the safety of the environment that includes substances which are toxic, infectious, mutagenic, carcinogenic, teratogenic, explosive, flammable, corrosive, oxidizing and radioactive" (p. 16) [32]. The worldwide generation of hazardous waste is reported by the Basel Convention[2] on the control of transboundary movements of hazardous wastes and their disposal for 2000 and 2001 to be 318 and 338 million tonnes, respectively [33]. These figures are as reported by the signatories to the Convention and are only estimates. Since one fourth of all countries, including the U.S., have not ratified the convention, the estimates are necessarily incomplete. Moreover, there are many inconsistencies in identifying and defining the different waste types, and even more complicated criteria for reporting the trade in those wastes. Hence, the precise amounts of toxic waste generated and traded internationally are not known.

---

[2]The provisions of the Convention center around three principal aims: (i) the reduction of hazardous waste generation and the promotion of environmentally sound management of hazardous wastes, wherever the place of disposal; (ii) the restriction of transboundary movements of hazardous wastes except where it is perceived to be in accordance with the principles of environmentally sound management; and (iii) a regulatory system applying to cases where transboundary movements are permissible.

Hazardous waste is directly associated with creating environmental and health concerns. Most of the pollutants are persistent in the environment once introduced and cause ecological destruction, and most importantly serious health issues for humans including cancers, respiratory diseases and immune and reproductive disorders. The tendency to cause serious environmental and health issues and the difficulty in managing these types of waste has made the disposal a grave socio-environmental concern. Hence, many countries and firms throughout the world are trying to pass their "headache" to other countries and communities. This has made hazardous waste management a global business carried out through many channels by legitimate or criminal means. These businesses have made their way to the global South mainly due to the lower costs and weak environmental regulations in developing countries [10].

Many conventions and treaties govern the international trade or the transboundary movement of hazardous wastes. These include, but are not limited to:

- the Basel Convention for the control of transboundary movements of hazardous wastes and their disposal;

- the 1991 Bamako Convention which prohibits the import of hazardous wastes to Africa and the control of transboundary movement of hazardous wastes within Africa [34];

- the 1996 Izmir Protocol, which prohibits the trade in toxic waste in the Mediterranean region [35];

- the 1995 Waigani Convention, which covers the South Pacific region [36];

- a European Union regulation, passed in 1997, prohibiting the export of hazardous waste to non-OECD countries.

Despite all the national and international regulations and a global treaty ratified by 175 countries, global trade in hazardous waste is continually taking place throughout the world. Basel Convention came into effect as a result of the illegal hazardous waste dumping activities occurred in Africa and other parts of the world in the late 1980s [37]. After almost three decades, movement of hazardous waste continues throughout the world. Based on import data submitted by its member countries to the Basel Convention in 2009, roughly 12.1 million metric tonnes of hazardous wastes were transported among all parties either for disposal, recovery, or unspecified reasons [39]. The U.S., the largest producer of hazardous waste in the world is yet to ratify the Basel Convention. There are still firms in the United States that have continued to export toxic waste to developing countries [9] although the convention states that a party shall not permit hazardous waste or other waste to be exported to a non-party or to be imported from a non-party [36]. Hence, the precise amounts of toxic waste generated and traded internationally are not known.

A tragedy in the low-income African country, Côte d'Ivoire, confirms that all parties involved will need to adhere to much stricter guidelines and enforcement of all national and international conventions on trade in hazardous waste. In 2006, a vessel named "Probo Koala" unloaded 500 tonnes of petrochemical waste into trucks, which ultimately dumped the waste around the residential areas in the largest city in Côte d'Ivoire. The incident, which took place in August 2006, killed at least 12 people, and well over 100,000 people had to seek medical attention [40].

The hazardous waste flow is usually from countries with high disposal costs and stricter regulations to countries with low disposal costs and weaker regulations. The key point ignored by many stakeholders involved is that developing countries are in any case less equipped to manage complex toxic waste [41]. The Basel convention highlights that transboundary movement of hazardous wastes to developing countries, many of which are incapable of handling such waste, do not constitute environmentally sound management as required by the Convention [37]. According to Basel Action Network [42] and Greenpeace [43], this ultimately leads to toxic waste shipped to developing countries being disposed of haphazardly without proper environmental measures.

The Basel Convention states that the parties should reduce the generation of hazardous waste to a minimum, taking into account social, technological and economic aspects [37]. It further establishes that waste should only be sent to another country if the exporting state does not have the capacity to dispose of the waste in an environmentally sound manner. The final criterion for international trade is that waste should constitute "raw materials" that can be used by the importing countries [37]. Increasing amounts of waste are exported for recycling, driven by the increasing prices of secondary raw materials. This criterion has become a major loophole in the convention, which has created a path for another global problem. Developed countries and firms, instead of exporting their waste for disposal have now started using a new path; exporting for recycling. These recycling practices are not always "green" and generate hazardous by-products that need to be disposed of again [18].

To put an end to this "pollution by recycling" business, an amendment to the convention was introduced in 1995. The amendment provides for the prohibition of exports of all hazardous wastes covered by the Convention that are intended for final disposal, reuse, recycling and recovery from countries listed in Annex VII,[3] to developing countries that are not [37, 44]. This amendment known as the Basel Ban Amendment, has not yet entered into global legal force for a lack of signatories. As of August 2010, it has been ratified by 69 countries and has been implemented by 34 of the 41 developed countries to which its export ban applies [18, 45]. The United States, New Zealand, Australia, Israel, Japan, South Korea, and Canada have not agreed to implement the Basel Ban Amendment in their countries [18].

Global conventions and laws do not govern the toxic waste trade between developing countries nor between developed countries [10]. For example, most US hazardous waste ends up in Canadian landfills. Roughly 500,000 tonnes of hazardous waste cross the Canada/U.S. border annually [46]. While the United States has banned the dumping of untreated hazardous waste in their landfills, Canada has allowed this practice and has opened its borders for hazardous waste from the United States [46].

---

[3]Parties and other States which are members of the OECD, EC, and Liechtenstein.

An alternative new trend has emerged in the world, where the polluting industries are relocating from their home countries to less resilient environments in the developing countries. Many multinational corporations in the chemicals industry relocated much of their production to developing regions during the 1990s. European chemical companies are being attracted to Asia, mainly because they were driven away from their home countries by high labor costs and strict regulations [47]. For example, Dr. Manfred Schneider, chief executive of Bayer in 1994, has stated that the main disadvantages for the chemical industry are the high labor costs, expensive social security systems, and the widespread environmental regulations imposed by the state [47].

When managing hazardous waste becomes a challenge in financial, social, and technical terms, countries and firms have to deal with their own waste in some way. If adequate waste management measures are not in place, firms will inevitably stockpile their waste, increasing the risk of environmental hazards. Few examples of stockpiled hazardous waste in Australia are given below.

1. Australia has total holdings of around 4,300 cubic meters of radioactive waste. The nuclear waste has resulted from three main sources—radioactive medical, scientific and industrial waste—spent nuclear fuel from Australia's research reactor at Lucas Heights near Sydney, and site contamination from British nuclear weapons tests conducted in South Australia in the 1950s [32].

2. Orica, a chemical manufacturing company, has stockpiled a large quantity of Hexachlorobenzene (HCB) at the Botany Industrial Park (BIP) in Sydney. HCB was generated as a waste by-product in solvent and plastic manufacturing plants at BIP between 1963 and 1991. Orica worked on applications to export the waste for destruction in high temperature incinerators in Europe between 2006 and 2010. However, in December 2010, the Danish Government made a decision to cancel its approval for the environmentally sound destruction of Orica's HCB waste at a treatment facility in Denmark. As of July 2011, the company was unsuccessful in finding a solution for destruction of the HCB waste one of the world's largest remaining HCB stockpiles [48].

3. The Fremantle Steam Laundry in Hamilton Hill burst into flames on May 13, 2010. Fire fighters had to evacuate nearby residents because the factory had a stockpile of the dry cleaning chemical perchloroethylene (PCE) [49].

4. According to the 2010 National Waste Report, Australia generates about 106,000 tonnes per year of e-waste in the form of old televisions and computers. A number of hazardous substances are found in these electronic wastes such as lead, mercury, cadmium, and arsenic [26].

It is important to note how stockpiled toxic waste will get disposed eventually as hazardous waste has a high tendency to find its way to a remote location or to a marginalized community within the same country or another, as the final resting place.

4.5   E-WASTE: SKELETONS OF MODERN TECHNOLOGY

"Machines that could, just months before, process a billion instructions per second, send a message clear around the world with the stroke of a key, or hold a library of books in a palm-sized drive - have found their end as metal and plastic skeletons, in the world's most sorrowfully poor communities to be subjected to hammer and fire, emitting deadly smoke and fume. Shouldn't there be a law?"(p. 1)

Jim Puckett [18]

The list of electronic goods that we use today, ranging from watches, mobile phones, laptops, televisions to refrigerators, is almost uncountable. We live in an era of information technology where messages are sent across the world in milliseconds and transactions are completed in the blink of an eye. Electronic usage has sky rocketed during the past two decades and consumer demand for such items is higher than for some of the common goods we consume every day. This demand and over consumption has created a long lasting global problem. As Puckett (2011) has stated, "they (electronics) satisfy our ego desires, our need for speed, our competitive edge, perhaps, but they also create mountains of a new type of waste all around the world: *e-waste*" (p. 2) [18].

According to the United Nations Environment Program, 20–50 million metric tonnes of e-waste is produced per year globally [50]. Demand is increasing rapidly as consumers acquire more and more electrical and electronic goods and frequently exchange appliances for new ones. For example, in the United States, computer ownership has increased from 1 per 1,000 in 1975 to roughly 1 per person in 2010 [18]. The rapid technology change, low initial cost, high obsolescence rate have resulted in a growing problem of waste electronics around the globe. The products are out of date within two to three years and give way to new brands and models. The old gadgets end up in second-hand markets or in waste dumps and landfills, adding to the remaining piles of e-waste.

The core problem with e-waste is the amount of toxic chemicals it contains, making e-waste one of the most universally available hazardous waste type. For example, a typical desktop computer contains metals such as copper, aluminium, lead, gold, zinc, nickel, tin, silver and iron along with platinum, palladium, mercury, cobalt, antimony, arsenic, barium, beryllium, cadmium, chromium, selenium, and gallium [51]. Most of these heavy metals are toxic even in very low concentrations that can cause threats to humans, animals and the environment.

Most of the electronic waste ends up in landfills [52]. According to Greenpeace (2009), in many European countries there are regulations to prevent electronic waste being dumped in landfills. [53]. However, the practice continues in many countries. To reduce e-waste going in to landfills, many multinational corporations in the electronics business are seen as actively promoting recycling in popular media, requesting their consumers to dispose e-waste only in designated collection points. Instead of manufacturing long-lasting electronic equipment and reducing the rate of obsolescence, manufacturing companies continue to assert that recycling is the only answer. Many people actively support e-waste recycling, however little is known about how and where these recycling activities ultimately take place.

Recycling e-waste can be an expensive process in developed countries with strict regulations to protect the environment and to ensure the health and safety of the workers [54]. Without compromising their profits, e-waste dealers in developed countries have chosen a more profitable option; sending e-waste to "far away" places in developing countries where no rule or law exists to control these detrimental practices [55]. Why is it that e-waste is shipped to low income countries to be smashed and burned by marginalized people struggling with their day-to-day lives? Why is it that they pay with their health and well-being for a waste generated from a product that they never consumed? Will the firms in the developed countries ready to pay compensation for the poor health conditions and low life expectancy in these communities caused by toxic waste? Unfortunately, no one seems to "bother" to find answers for any of these issues.

Recycling e-waste in many developing countries is not carried out in a socially or an environmentally responsible manner [18]. Much of the e-waste is not actually processed but is instead dumped in local villages and water sources [56]. People who are working in these dumps usually smash and burn the electronic waste products, such as televisions, laptops and computers, to recover small amounts of valuable metals [18]. Often recovery activities are performed bare handed without any protective gear or equipment. Such illegal e-waste dumpsites have been documented in Lagos in Nigeria, Accra in Ghana, Guiyu in China, Karachi in Pakistan, and Delhi in India [55].

U.S. is one of the largest e-waste generators of the world [55]. In 2005, more than 75% of Cathode Ray Tube (CRT) containing products collected in the United States for recycling primarily were exported to developing countries [57]. As indicated previously, the United States has not ratified the Basel Convention and can continue to externalize the risk by exporting their e-wastes to developing nations. It was reported that reprocessors from China, Malaysia, India, Vietnam, Hong Kong, Taiwan, Pakistan, Egypt, and Sri Lanka made the highest number of requests to purchase CRTs on two US internet E-commerce web sites from February to May 2008 [58]. In these countries the demand for raw materials are high, thus high profits can be earned by selling extracted materials. Also, in some countries, there is a good market for equipment that can be repaired easily and traded in the second hand market. These factors have largely contributed to set up e-waste recycling plants in many developing countries. The facilities attract a large pool of unemployed individuals willing to remove metals from e-waste with no worker protection. The workers are paid extremely low wages while middle men who manage these facilities earn high profits [9].

At this very moment, many of us are comfortably sitting in front of our televisions, monitors, and laptops watching a program, tapping a keyboard or reading an article. At this very moment, thousands of miles away somewhere in a wasteland, people like us are dealing with the tonnes of electronic waste we have created. As we come to the end of this section, we have to bear in mind that this "away" is a real place; in Guiyu in China, Karachchi in Pakistan, or Lagos in Nigeria, where people like us are struggling with our waste for their daily survival.

REFERENCES

[1] Clapp, J., The distancing of waste: Overconsumption in a global economy, in *Confronting Consumption*, T. Princen, M. Maniates, and K. Conca, Editors. 2002, MIT Press: Cambridge. p. 155–176. 69, 72

[2] Daskalopoulos, E., O. Badr, and S.D. Probert, Municipal solid waste: a prediction methodology for the generation rate and composition in the European Union countries and the United States of America. Resources, Conservation and Recycling, 1998. **24**(2): p. 155–166. DOI: 10.1016/S0921-3449(98)00032-9 69

[3] Hoornweg, D. and P. Bhada-Tata, What a waste: a global review of solid waste management, in *Urban development series; knowledge papers no. 15*. 2012, The Worldbank: Washington D.C. 69

[4] UN-HABITAT, Solid Waste Management in the World's Cities: Pre-publication presentation, D.C. Wilson, A. Scheinberg, and L. Rodic-Wiersma, Editors. 2009, UNON Print Shop: Nairobi. 69, 72

[5] Scheinberg, A., A Bird in the Hand: Solid Waste Modernisation, Recycling and the Informal Sector, in *Solid Waste Planning in the Real World, CWG-Green Partners Workshop 2008*, Cluj, Romania. 69, 72

[6] Bullard, R.D., Solid Waste Sites and the Black Houston Community*. *Sociological Inquiry*, 1983. **53**(2–3): p. 273-288. DOI: 10.1111/j.1475-682X.1983.tb00037.x 70, 74

[7] Pastor, M., J. Sadd, and J. Hipp, Which Came First? Toxic Facilities, Minority Move-In, and Environmental Justice. *Journal of Urban Affairs*, 2001. **23**(1): p. 1–21. DOI: 10.1111/0735-2166.00072 70, 72, 74

[8] Rothman, D.S., Environmental Kuznets curves—real progress or passing the buck?: A case for consumption-based approaches. *Ecological Economics*, 1998. **25**(2): p. 177–194. DOI: 10.1016/S0921-8009(97)00179-1 70, 73

[9] Gibbs, C., E. McGarrell, F., and M. Axelrod, Transnational white-collar crime and risk : Lessons from the global trade in electronic waste. *Criminology and Public Policy*, 2010. **9**(3): p. 543–560. DOI: 10.1111/j.1745-9133.2010.00649.x 70, 77, 81

[10] Clapp, J., Seeping Through the Regulatory Cracks. *SAIS Review*, 2002. **22**(1): p. 141–155. DOI: 10.1353/sais.2002.0004 70, 77, 78

[11] Rothman, D. S., Environmental Kuznets curves—real progress or passing the buck? A case for consumption-based approaches. *Ecological Economics*, 1998. **25**: p. 177–194. DOI: 10.1016/S0921-8009(97)00179-1 71

[12] Hamilton, C., R. Denniss, and D. Baker, *Wasteful Consumption in Australia*, 2005, The Australia Institute: Australia. 71

[13] The World Bank, *What a Waste : Solid Waste Management in Asia*, 1999, Urban Development Sector Unit, East Asia and Pacific Region. 71

[14] United Nations Environment Programme, *Solid Waste Management*, 2005, United Nations Environment Programme. 71

[15] Medina, M., Waste Picker Cooperatives in Developing Countries, in *Wiego/Cornell/SEWA Conference on Membership-based Organisations of the Poor* 2005: Ahmedabàd, India. 72

[16] Widmer, R. et al., Global perspectives on e-waste. *Environmental Impact Assessment Review,* 2005. **25**(5): p. 436–458. DOI: 10.1016/j.eiar.2005.04.001 72

[17] Clapp, J. and T. Princen, Out of sight, Out of mind: Cross border traffic in waste obscures the problem of consumption. *Alternatives journal,* 2003. **29**(3): p. 39–40. 72

[18] Puckett, J., *A place called away.* 2011. 72, 76, 78, 80, 81

[19] Bullard, R.D., Environmental Justice in the 21st Century: Race Still Matters. 2001. **49**(3/4): p. 151–171. 73, 74

[20] Pellow, D.N., The Politics of Illegal Dumping: An Environmental Justice Framework. *Qualitative Sociology,* 2004. **27**(4): p. 511–525. DOI: 10.1023/B:QUAS.0000049245.55208.4b 73, 74, 75

[21] Gawande, K. et al., Internal migration and the environmental Kuznets curve for US hazardous waste sites. *Ecological Economics,* 2000. **33**(1): p. 151–166.
DOI: 10.1016/S0921-8009(99)00132-9 73

[22] Birdsall, N. and D. Wheeler, Trade Policy and Industrial Pollution in Latin America: Where Are the Pollution Havens? *The Journal of Environment and Development,* 1993. **2**(1): p. 137–149. DOI: 10.1177/107049659300200107 73

[23] Faber, D. and E. Krieg, *Unequal exposure to ecological hazards: Environmental injustices in the Commonwealth of Massachusetts.*, 2001, Northeastern University: Boston, MA. 74

[24] Robinson, D. M., *Environmental Racism: Old Wine in a New Bottle.* ECHOES, 2000. 74

[25] Australian Bureau of Statistics, Regional Population Growth, Australia, 2007–08, *Australian Bureau of Statistics,* 2009. 75

[26] Environment Protection and Heritage Council, *National Waste Report 2010,* Department of the Environment Water Heritage and the Arts, 2010, NEPC Service Corporation. 75, 76, 79

[27] LMS Generation Pty Ltd, Submission in response to A National Waste Policy : Consultation Paper, 2009. 75

[28] Wright Corporate Strategy PTY LTD., Review of the application of landfill standards, 2010, *Department of the Environment Water Heritage and the Arts.* 75

[29] Waste Management Association of Australia, *National Landfill Survey 2007–2008 results,* 2009. 75

[30] South East Resource Recovery Regional Organisation of Councils, Submission in response to A National Waste Policy: Consultation Paper, 2009. p. 8. 76

[31] City of Darebin, Submission in response to A National Waste Policy: Consultation Paper, 2009. p. 2. 76

[32] Environmental Defenders Office WA (Inc), Waste Management in Western Australia: Current Law and Practice and Recommendations for Reform, 2007, Environmental Defenders Office WA (Inc),. 76, 79

[33] Basel Convention, *The Basel Convention at a Glance,* 2007, UNEP/SBC, International Environment House: Switzerland. 76

[34] Bamako Convention, Bamako Convention on the Ban of the Import Into Africa and the Control of Transboundary Movement and Management of Hazardous Wastes Within Africa, *Organization of African Unity,* 1991. 77

[35] Izmir Protocol, Protocol on the Prevention of Pollution of the Mediterranean Sea by Transboundary Movements of Hazardous Wastes and their Disposal, 1996. 77

[36] Secretariat for the Waigani Convention, The Waigani Convention to Ban the Importation into Forum Countries of Hazardous and Radioactive Wastes and to Control the Transboundary Movements and Management of Hazardous Wastes within the South Pacific Region, 1995. 77

[37] Secretariat for the Basel Convention, Basel Convention : On the control of transboundary movements of Hazardous wastes and their disposal, UNEP, 2011: Switzerland. 77, 78

[38] Secretariat for the Basel Convention, *Illegal traffic under the Basel Convention,* UNEP, 2010.

[39] Secretariat for the Basel Convention, Basel Convention reporting database, Summary tables on Trasboundary Movements among reporting Parties (in metric tons), 2009 Summary Disposal/Recovery/Unespecified, 2009, International Environment House,Geneva, Switzerland. 77

[40] Widawsky, L., In My Backyard: How Enabling Hazardous Waste Trade to Developing Nations Can Improve the Basel Convention's Ability to Achieve Environmental Justice. *Environmental Law,* 2008 **38**(2): p. 577. 78

[41] European Environment Agency, *Waste without borders in the EU? Transboundary shipments of waste,* 2009, EEA: Copenhagen, 2009. 78

[42] Basel Action Network. *Toxic Trade News* 2011 Available from: `http://www.ban.org/toxic-trade-news/`. 78

[43] Greenpeace International, Chemical contamination at e-waste recycling and disposal sites in Accra and Korforidua, Ghana, in Technical Note 10/2008, August 2008, Greenpeace Research Laboratories. 78

[44] Puckett, J., *The Basel Treaty's Ban on Hazardous Waste Exports: An Unfinished Success Story. International Environmental Reporter,* 2000. **23 INER 984**. 78

[45] Basel Action Network, *Country Status/Waste Trade Ban Agreements,* 2010, Basel Action Network. 78

[46] Environmental News Service, *Canadian Imports of U.S. Hazwaste Down in 2000.* 3 August 2001. 78

[47] Abrahams, P., The Dye is Cast by Growth and Costs, in 31 May 1994, *The Financial Times Limited.* 79

[48] Orica Ltd. Repackaging and Destruction of Stored HCB Waste Project. Botany Transformation Projects 26 July 2011 26 July 2011; Available from: `http://www.oricabotanytransformation.com/?page=14andproject=83`. 79

[49] Wainwright, S. Toxic Fire Exposes Working Class Community. Green Left Weekly, 15 May 2010, 2010. 79

[50] United Nations Environment Programme. *Waste.* Available from: `http://www.unep.org/climateneutral/Topics/Waste/tabid/156/Default.aspx`. DOI: 10.1080/00185868.1994.9948474 80

[51] Grossman, E., *High Tech Trash: Digital Devices, Hidden Toxics, and Human Health,* 2007, Island Press. 80

[52] Clean Up Australia, E-Waste Fact Sheet, November 2009, Clean Up Australia, NSW Australia. 80

[53] Greenpeace Inetrnational. *The e-waste problem.* 2005; Available from: `http://www.greenpeace.org/international/en/campaigns/toxics/electronics/the-e-waste-problem/`. 80

86 REFERENCES

[54] Ongondo, F.O., I.D. Williams, and T.J. Cherrett, How are WEEE doing? A global review of the management of electrical and electronic wastes. *Waste Management*, 2011. **31**(4): p. 714–730. DOI: 10.1016/j.wasman.2010.10.023 81

[55] Greenpeace Inetrnational. Where does e-waste end up? 2009; Available from: `http://www.greenpeace.org/international/en/campaigns/toxics/electronics/the-e-waste-problem/where-does-e-waste-end-up/`. 81

[56] Pellow, D.N., *Resisting Global Toxics: Transnational Movements for Environmental Justice*, 2007, MIT Press. 81

[57] U.S. Environmental Protection Agency (EPA), *Management of electronic waste in the United States*, Environmental Protection Agency (EPA), 2007, US. 81

[58] U.S. Government Accountability Office (GAO), Electronic waste: EPA needs to better control harmful U.S. exports through stronger enforcement and more comprehensive regulation, 2008, United States Government Accountability Office: US. 81

CHAPTER 5

Waste Management in the Global South: A Sri Lankan Case Study

Randika Jayasinghe

"The world has enough for everybody's need but not enough for everybody's greed."

Mohandas Gandhi (October 2, 1869 - January 30, 1948)

5.1 INTRODUCTION

In a world dominated by profit-driven organizations, money-making machines, and investment opportunities, waste management has become a grave environmental issue. Increasing waste generation and poor waste management has created many environmental, economic and public health problems in the world, particularly in developing countries. The urban, agricultural, and industrial activities in the developing countries are expanding, spurred by increasing populations. Due to rapidly growing consumer societies and expansive growth policies during the last few decades, the lifestyles and consumption trends in the global South have changed dramatically.

New lifestyles associated with things like fast food outlets, mobile phones, soft drinks, and disposable items are creating more waste, causing significant pressure on both socio-economic and environmental well-being. Many multinational companies have expanded their products and services to the South. We can see more fast-food outlets, more single-serving beverages and more packaged food on the shelves than ever before. Many profit-oriented companies hardly consider the potential waste management problems that go hand-in-hand with changing lifestyles and consumption patterns [1].

To give a good example, the President of Coca Cola, Donald R. Keough, was quoted as saying [2],

"When I think of Indonesia – a country on the Equator with 180 million people, a median age of 18, and a Moslem ban on alcohol – I feel I know what heaven looks like."(p. 39)

Our focus is not on globalization and multinational companies. Nevertheless, this quote highlights the effect of globalization on the South. The global South has a population of more than

5 billion people [3]. If, on average, one beverage was consumed per person per month, then 60 billion cans and bottles would be added to the waste stream every year. Will it create a "heaven" for the people who live in developing countries with more waste to collect and limited resources to spend on waste management? Are the manufacturing organizations willing to clean up the waste and/or allocate funds for waste management in developing countries? We need to ask "Who benefits and who eventually pays the cost?." This chapter attempts to address some of these issues with regard to waste management in the global South triggered by rapidly changing consumption patterns.

Although much literature deals with a variety of topics in the field of waste management, the waste problem remains unchanged. This is not surprising, as most concepts that dominate popular thinking, and also academic work, are based on the ideas of industrialized nations and highlight knowledge and technology suitable for those countries, but not for the developing world. Only a few have been published that provide knowledge and information required by developing countries. Keeping this in mind, we have dedicated this chapter to discuss waste management in the developing world through a social and environmental justice lens. We will continuously raise the question "Who benefits and who pays; who wins and who loses?" in this never-ending battle with waste.

5.2   WASTE MANAGEMENT IN THE GLOBAL SOUTH

"Weighed down by the heavy rains of two typhoons that hit the Philippines in 2000, a monstrous mountain of garbage collapsed on a cluster of shanties outside Manila. The accident claimed more than 200 lives in the small community known as Lupang Pangkao, or the Promised Land."

S. Kriner (DisasterRelief.org) [4]

This is not the only garbage dump landslide that claimed lives in a developing city. The same story is continually repeated for people of poor communities living in close proximity to mountainous garbage dumps. The following table (Table 5.1) presents few reported large garbage landslide events, which have occurred between 2000-2008 and the resulting death toll. People who died in these unfortunate accidents were not rich people, but people who were living in extreme poverty; people who either lived in close proximity to the city's garbage mound or who were scavenging in these dump sites to supplement their daily incomes [5].

As mentioned earlier, there is a trend of rapid urbanization and industrial development driven by globalization in developing countries. One of the most obvious impacts of this transformation can be seen in the form of unattended piles of waste in many developing countries. Waste is a serious multi-faceted problem in countries with a poor economic status. Waste, which is often mixed with human and animal excreta, is dumped haphazardly in the streets and in drains, reducing the aesthetic value of the environment, causing water pollution, reducing soil fertility, contributing to flash floods, providing breeding grounds for insect and rodent vectors and spreading diseases. Managing the problem is difficult due to scarce resources available. Regrettably, the poor in the society suffer most from the disorganized waste management as the authorities tend to allocate their limited waste

Table 5.1: Garbage landslides and the resulting death toll (Source: The Landslide Blog)

Date	Location	Country	Death toll
10/07/2000	Manilla	Philippines	287
7/10/2005	Bello, Medellin	Colombia	43
1/09/2005	Padang, Sumatra	Indonesia	25
21/02/2005	Cimahi, Bandung	Indonesia	143
20/06/2008	Guatemala City	Guatemala	50

management resources to the affluent areas where citizens with more financial and political power reside [6]. There is more to the problem as rapid growth of shantytowns on the outskirts of cities pose new infrastructure challenges for the authorities.

It is important to discuss briefly the waste management trends in developing countries as a foretaste to our whole discussion. The waste management in the global South is very complicated and characterized by different factors such as [1, 6, 7]:

- low labor costs;

- lack of funds and poorly distributed capital and physical infrastructure;

- a waste stream dominated by organic waste often mixed with all types of other waste including industrial hazardous wastes [8];

- a complex informal sector that is very active in the collection, separation, and recycling of waste;

- inadequate governance and weak enforcement of legislation;

- poor public involvement; and

- lack of appropriate planning.

Generally, the wastes generated in developing countries have a higher percentage (50%–80%) of organic matter in the waste stream due to the consumption of fresh fruits, vegetables, and unpackaged food [7, 9]. This is an important consideration, not only because it constitutes a substantial portion of the waste stream, but also because it is unseparated and can have potentially adverse impacts on the environment including leachate generation and methane emission. It is equally important to note that, changes in consumption patterns have caused tremendous alterations in the waste composition [10]. The waste management problem has become more complicated with new trends which have emerged over the past few years. Plastics, paper, and electronic wastes have made their way into the waste stream due to the rapid economic expansion in developing countries [1].

Accumulation of these non-degradable wastes pose new challenges for waste management in developing countries.

The differences between waste quantities and compositions in developing countries persist from the national level to the community level within the same country. Waste generation rates and compositions can vary based on [1, 11]:

- the socio-economic development;

- degree of industrialization;

- culture, consumption patterns, and lifestyles; and

- and climatic conditions.

However, a more generalized assumption would be, the greater the economic development rate and the growth of the urban population, the greater the amount of waste produced [1] and shift in composition to include more packaging in the waste stream.

Waste generation in Asia is estimated to have reached 1 million tons per day in 2005 [8]. A World Bank study states that urban areas in Asia spent USD 25 billion on solid waste management in 2005, and that this figure will increase to USD 50 billion in 2025 [1]. It will undoubtedly be interesting to find out how these funds are allocated and spent on waste management in developing countries. It is important to note that waste management costs in developing countries are high and the level of service is low [12]. Waste management costs consume 20–50% of municipal revenues, but only 50–70% of residents will receive the service with overall collection service levels remaining low [13]. The 30–50% who receive the poorest waste collection service includes the low income neighborhoods, slums and squatter settlements [9] with very infrequent collection levels (often less than once a every two weeks).

Final disposal in many developing countries means merely transporting the collected waste to an available open space. Open dumping of wastes in barren lands, wetlands, water bodies, and beside roads are still the most predominant final disposal method. This is followed by burning and/or burying of waste to reduce the volume and exposure to the environment [14]. Not all developing countries have properly maintained sanitary or engineered landfills [8]. Often the authorities and waste management companies misuse these concepts where waste is dumped in an open land and a soil cover is put on top of the waste materials. These dumpsites have no leachate or gas collection systems in place and are often overflowed. Composting is another preferred method, mainly due to the high percentage of organic material in waste. This method is practiced widely but has failed to achieve satisfactory results on a larger scale, mainly due to poor involvement of the public and lack of separation of organic and inorganic waste at the source [7].

As mentioned earlier, often the collected waste is dumped in open dumpsites or in poorly managed landfills in close proximity to where poor people reside [15]. This is certainly not the solution, in fact it shifts the problem from one place to another and creates far more complicated problems. It is not the shifting of the problem (that happens the world over), but to shifting to people who are ill-equipped to deal with the health issues, and who put themselves in risk by scavenging, creates the greatest issues. However, most authorities seem to be continuing this practice contentedly as poor communities rarely raise their voices against the authorities. Even if they make a complaint, the probability of them not being heard by the relevant authorities is high. This raises a question: Would the authorities take no action and turn a blind eye if an influential person such as a politician or a businessman made a request to remove the waste from their vicinity? This discrimination is one of the main issues of a socially just waste management system in developing countries.

Another important strategy of waste management is the strong participation of informal sectors in waste management. Most developing countries do not pay much attention to recycling or recovery programs as they are preoccupied with waste collection and disposal [9]. Hence, the recycling programs largely rely on the informal sector, commonly known as "scavengers" or "waste pickers." These informal sectors are comprised of poor and marginalized groups in society who have resort to scavenging to supplement their income and even for their daily survival [16]. Such work is generally labor-intensive, unsafe, and generates very low income [9, 10]. Recovered and recyclable products are sold to dealers, or processing facilities before they are finally sold to manufacturing enterprises. Waste pickers are in the lowest level of the waste management sector, often exploited and paid very low prices for the collected materials by dealers or middlemen [16]. The middlemen enjoy the actual benefit, earning high profit margins by selling the recyclables to manufacturers.

It is clear that unless effective waste management measures are introduced and effectively enforced, the burden of solid waste management in developing countries will be worsened given their rapidly growing consumer societies. However, to find solutions, firstly we need to understand the following related to waste management in a developing country.

- What does "waste management" actually mean for a country in the global South?

- Does it mean changing their consumption patterns and lifestyles?

- Can developing countries follow the footsteps of developed countries and adopt their technologies to solve the problem?

- Does it mean ignoring the local circumstances, culture, stakeholders, and local knowledge in developing countries?

- Can such an approach be effective, let alone sustainable in the long run?

Understanding these questions demonstrates that waste management in developing countries is interconnected with its socio-economic context and requires an approach different to that of industrialized countries to be effective and sustainable.

Rather than developing a general discussion on the social issues of waste management in developing countries, the following sections will focus on a specific scenario. We will pay particular attention to waste management in Sri Lanka; a small developing country with rapid economic growth and a high population density. We understand that waste management greatly differs from one country to another based on a number of factors discussed earlier. It varies from city to city in the same country and even within the communities in the same city. However, we hope that this discussion on Sri Lanka's waste management will provide greater insights into the waste management in a developing country from a social and environmental justice viewpoint.

(Randika Jayasinghe, the author of this chapter has worked and carried out many research studies in the waste management sector in Sri Lanka. Hence, unless otherwise stated, data and statistics related to waste management are the author's own research findings collected through surveys, interviews, focal group discussions, and observations.)

5.3   SRI LANKA: THE PEARL OF THE INDIAN OCEAN

"Dear me, it is beautiful…sumptuously tropical—a dream of fairyland and paradise."

Mark Twain, *when his ship reached Ceylon [Sri Lanka] in the 1890s.*

Sri Lanka is an island nation in the Indian Ocean located to the south of the Indian subcontinent. It has many names. The Chinese called it the *Land without Sorrow*. The Arabs called it *Serendip*. The Hindu epic *Ramayana* referred to it as the *Garden in the Sky*. It is popularly referred to as the "Pearl of the Indian Ocean," a name which reveals the beauty and prosperity of the island. Marco Polo (1254-1324) described Sri Lanka as "the finest island of its size in the whole world" [17]. The island is full of culture and colour, blessed with nature's gifts such as sandy beaches, mountains, forests, flowers and wildlife that Mark Twain described as, "all harmonious, all in perfect taste."

The island has a total land area of 65,610 km^2 with a population of 20.4 million inhabitants. This makes the country a densely populated island in the world with a population density of 308/km^2 [18]. Sri Lanka's economy was entirely based on agriculture and trade for many centuries. However, the country began to shift away from a socialist orientation in 1977 to an open economy. Since then, the government has been privatizing, and opening the economy to international competition making tourism, tea export, apparel, rice production, and other agricultural products the main economic sectors of the country. According to recent IMF and World Bank estimates, Sri Lanka's economy is one of the strongest in the South Asian region with an economic growth rate of 3.5% in 2011 [18].

Before this, the economy of the country had been severely affected by the 2004 Indian Ocean Tsunami and a number of rebellions, such as the 1971, the 1987–1989 and the 1983-2009 civil wars. In particular, the civil war which lasted for almost 26 years had devastating effects on the economy and development as a large portion of the financial budget was allocated for war and security-related expenses. For this reason and the government's attempt to accelerate the pace of industrial development, the country has failed to pay attention to one of the key necessities of the society; the management of waste [19].

Poor waste management has incurred severe penalties and adverse impacts on the environment and on public well-being. It has significantly affected the aesthetics, local traffic, flooding, air, and water quality. It has also increased the incidence of diseases born by vectors such as rodents, mosquitoes and flies. Hence, the waste problem has become a grave socio-environmental concern, particularly in the urbanized areas of Sri Lanka with its high population density and lack of suitable land for waste disposal.

The population growth, changing consumption patterns, improved living standards, and rapid urbanization have increased the waste generation in the country over the years [7]. With the ever-increasing population, the waste problem has grown from being an annoyance for Regional and Local Governments to now being an issue of national importance. Haphazard waste disposal has been identified as a major factor contributing to environmental degradation in the National Policy on Solid Waste Management [20]. However, open dumping and open burning are still the most common final disposal methods practiced in the country [21]. The problem is most severe in the Western Province, the most densely populated area in Sri Lanka [22] with a population density of $1445.5/km^2$ and with a population increase of 32.5% annually [23].

The total municipal solid waste generation in Sri Lanka is assumed to be around 6,400 tons per day with a per capita per day waste generation average of 0.85 kg in Colombo Municipal Council, 0.75 kg in other Municipal Councils, 0.60 kg in Urban Councils and 0.40 kg in Pradeshiya Sabhas—the smallest local administrative unit [7]. Daily waste collection in Sri Lanka is approximately 2,900 tons, of which 59% comes from the Western Province. This means that less than 50% of the solid waste generated each day gets collected. The overall coverage of solid waste collection services is low among most local authorities, averaging 65% for Urban Councils and just 50% for Pradeshiya Sabhas [24]. The bulk of rural waste is organic and biodegradable, and householders themselves are largely managing disposal [15]. However, this does not mean that waste management in semi-urban and rural areas should be taken lightly given that urbanization is occurring at a rapid pace.

The majority of funds for waste management from the local authorities are allocated for waste collection and transportation rather than for waste disposal and treatment [21]. All Local Authorities (LAs) spend more than 80% of their waste management budget in collection and transportation of waste, of which a significant amount is spent on salaries, allowances, maintenance and fuel costs [7, 8]. An insufficient amount is allocated for disposal, which leads to the wastes being dumped haphazardly without proper spreading, compacting and covering daily.

Sri Lanka does not have a state-of-the-art fully controlled sanitary landfill in operation. The first ever sanitary landfill site is still under construction at Dompe and completed the first phase in June 2012 [25]. Due to lack of proper disposal sites, most local authorities dispose waste at a number of uncontrolled open dumpsites. A majority of open dumps are in the low-lying areas, marshlands and abandoned lands that are filled with solid waste, predominantly as a land reclamation measure. The problem has become severe as wastes are not segregated and usually every type of waste such as industrial, slaughterhouse and hospital wastes are dumped in the same dumpsite, together with municipal solid waste [7, 13]. Lack of separation at the source has allowed infectious and hazardous wastes to enter the municipal waste stream, thereby to be exposed to the environment.

Proper waste management is crucial for the health of humans, animals and the environment as a whole. However, this is far from reality for Sri Lanka, where disorganized and unsafe waste management has triggered many socio-economic and environmental concerns. The haphazard disposal of waste has reduced the aesthetic value of cities, contributed to flash floods due to blockage of the drainage systems, streams and lagoons, and has increased health risks by providing breeding grounds for disease spreading vectors such as mosquitoes and rats. In most local authorities in Sri Lanka, the lack of environmentally acceptable solid waste management has created significant public health risks, particularly to the poor who live in close proximity to waste dumps and make their living through scavenging activities to supplement their unsatisfactory earnings [15].

At one time, the paradise of the Indian Ocean, Sri Lanka is now struggling with its mounting garbage dumps. As mentioned earlier, the scope of this book is to look at waste management through a different lens. Our aim is to see who is actually paying the price of poor waste management and who is benefiting from the services. In the following sections of this chapter, we will enter into the complex maze of waste management in Sri Lanka. We will try to find our way through this complicated system by using the lens of social and environmental justice.

5.4   WASTE MANAGEMENT THROUGH A SOCIAL AND ENVIRONMENTAL JUSTICE LENS

"The Not-In-My-Backyard (NIMBY) syndrome with respect to the siting of waste dumps has meant that some communities keep dumpsites out of their neighborhood while others are paid to take them."

Jennifer Clapp [26]

We generally tend to look at waste through a consumption lens. We often discuss the issues of high consumption, waste characteristics, compositions, and the economics of disposing wastes. It is even more common for us to look at waste through an engineering lens; an engineering challenge to be managed through landfill design and operation. Undeniably, it is important to consider all these aspects to develop a proper waste management system. Yet, considering only financial, economic, and technological aspects is like looking at the tip of the iceberg and trying to analyze the impact. The real problem is much greater and lies beneath what one can see. Looking at the waste problem through a social and environmental justice lens helps to explore these deeper concerns related to waste management in developing countries like Sri Lanka. It gets at questions that challenge the practices of contemporary consumer society itself: Who creates most waste? Who collects waste? Where is all this waste finally dumped? Why is waste dumped in those places but not elsewhere? Who pays the price of social costs associated with waste management?

Comprehensive studies on the relationship between waste generation rates and socio-economic parameters have been conducted in many developed countries. Such studies are few in developing countries and extremely rare in the Sri Lankan context. According to one of the studies carried out in the Western Province, the generation of organic waste per household increases with increases in property assessment tax value or income level. This is explained by the high food consumption trends of higher income groups [21]. People with high income, and thus high consumption patterns, create more waste compared to average and low-income people. People who live in marginalized conditions and with very little income create the least amount of waste. Yet, they are the group who are forced to deal with the waste that everyone else creates. Council waste collectors, street sweepers, informal waste collectors, and even waste collectors who are working for private companies are the poorest of the poor in the country; those at the lowest level of the waste economy who have very few alternatives as their livelihoods. According to a survey carried out in the Western province in 2011, some of these groups are earning very little income as low as 100-200 Sri Lankan rupees per day [27].

As with many developing countries, economic inequalities and power differences play a key role when it comes to waste collection in Sri Lanka. Most high-income people do not handle waste beyond their own garbage, which is conveniently taken away from their neighborhoods. According to a household survey conducted in the municipality of Moratuwa, municipal waste collection is available to only 56% of the households. The survey revealed that a high percentage of households from high and upper-middle income groups enjoy municipal waste collection services, while only a low percentage from the low-income groups have this service [21]. This situation prevails throughout the country and eventually has led the low-income and lower-middle income households to dispose of their waste along roads and any open space available.

The collection of waste seems to be the easy half of the problem. Disposal of what has been collected is what the authorities consider to be the real problem due to a lack of available sites. Most people would not even know, or care to know, where or how the waste they generate is disposed.

Once waste is out of sight people tend to think that it is someone else's responsibility. This *"someone"* is typically the local authorities, informal waste collectors and the private waste collecting firms. No one is ready to take responsibility for the waste they have generated and to bring about a lasting solution to the garbage menace. As stated earlier, the most common method of waste disposal in the country is littering and open dumping, the careless dropping of refuse or waste material in public places. Waste is disposed of on the roadside, in parks and in any convenient open space and outside houses (often in front of someone else's house). Some do not keep waste to be collected by the local authorities, but would rather go in the night when it is dark and throw the garbage bags in some place convenient to them. Moreover, some even throw their garbage bags out of the windows of moving cars.

This is largely because many people are making wrong assumptions about waste management. One such common belief is that it is acceptable to dump waste *"somewhere"* as it is the responsibility of the local authority to collect the waste and dump in a proper place. Often, people think that there are enough places for waste to go which creates an unfair situation as the poor have to tolerate the burden of waste most of the time. In short, *"Out of sight, out of mind"* seems to be the commonly accepted theory of waste disposal, even though someone else in a much more disadvantaged situation in the society has to pay the price of dealing with that waste.

The poor are often the first to carry the burden of environment degradation. They are the ones who suffer most from the improper waste management in the country, as waste collection in areas where poor people live are not adequate and often the city dump sites are located in close proximity to these areas [15]. Usually, every type of waste is dumped in these waste disposal sites without any precautionary measures to minimize environmental and health risks. Soil covering is rarely undertaken as it depends whether there is a strong public opposition to this method [7].

From a more general viewpoint, the perceived problem associated with waste is the offensive smell it creates and the piles of unattended waste that reduce the aesthetic beauty of an area. This is considered as a threat to the "status" of a neighborhood, which is why many would even pay to get the garbage removed from their immediate surroundings. However, the most important concerns are the socio-economic and environmental issues triggered by poor waste management. In particular, environmental problems caused by direct waste disposal and its consequences are widespread. This includes surface and ground water contamination and soil pollution through leachate, and air pollution by the burning of wastes. Improper waste disposal provides breeding grounds for disease spreading vectors such as rats and mosquitoes, releases greenhouse gases (GHG) by anaerobic decomposition of waste and impedes water flows in drainage channels which causes flash floods. Furthermore, dumping of industrial, slaughterhouse, and hospital wastes in open dumpsites and prolonged exposure to noxious gases by the burning of waste have caused far greater social and health problems for communities living close to the dumpsites.

Previous studies highlight that health issues in the marginalized communities are high due to poor waste management [8]. The waste composition in Sri Lanka is high in organic matter, which provides perfect habitats for rats that live on food remains in the waste, and which spread diseases mostly through their urine and faeces. The potentially fatal bacterial disease Leptospirosis is spread through human contact with rat urine, and occurs frequently near waste dumpsites [28]. The most widespread vector spread health issue related to uncontrolled waste dumps is mosquito-related diseases such as Malaria, Dengue, Japanese Encephalitis, and Filiarisis [29, 30]. For the nine months to the end of September 2011, there were 17,710 reported cases and 126 deaths of Dengue. Most of the Dengue patients and deaths were reported from the Western Province [31]. Dumpsites and waste piles provide an ideal breeding habitat for mosquitoes, particularly after rainy seasons. To curtail this problem and to extend the life of the dumpsites, local authorities regularly burn the waste, regardless of the resulting air pollution. This again causes neighboring communities to suffer from fumes, bad odor and smoke that can lead to respiratory and breathing related health problems.

As mentioned earlier, open dumping is a popular method of waste disposal. When a proper place is not found to cater for the purpose, waste is often dumped in slopes, waterways, and reservoirs, contaminating water sources. A good example is the Kelani River, one of the main sources of drinking water for the Colombo area, which is polluted by leachate from a number of dumpsites along the banks of the river [32]. Potable water is treated before distribution, and anyone who can afford to pay a monthly fee can obtain clean, pipe-born water. People who directly use the open natural water sources for their needs are the poor in the community, who cannot afford to pay for pipe-born water or any other safe water source. They are often left with no alternative but to consume water directly from polluted water sources.

When we look through a social and environmental justice lens, it is evident that marginalized people pay the cost of improper waste management. There is a timely concern to take measures to limit environmental degradation, as it will benefit everyone in the long run, but the marginalized communities the most.

In the next three sections, we explore the complex relationships of different parties involved in waste management; the politics of waste management, private sector participation, and the role of the informal sector.

5.5   SOLID WASTE GOVERNANCE: WHO IS RESPONSIBLE?

"Solid waste management and good governance are two sides of the same coin."

British scholar, Sir David Wilson

As with any developing country, waste management in Sri Lanka is a complex system with many stakeholders playing different parts throughout the process. It resembles a jigsaw puzzle with many parts missing; hence, the overall picture is still incomplete. In this section, we focus on the countless pieces of politics that belongs to this waste puzzle. We raise the following questions.

- Who are the stakeholders of waste management?

- Whose responsibility is waste? Who manages waste?

- Who is on top of the waste economy?

We all are part of the waste management system. We create waste and struggle to manage it, thus making each one of us "*waste stakeholders*" at some point, although we would not normally see ourselves as such. The main "recognized" stakeholders of waste management include the Environmental ministry, Central Environment Authority (CEA), local authorities, national government institutions, private companies working under contract to the local authorities, environment groups, and Non-Governmental Organizations (NGOs). They exercise power and authority in taking decisions related to waste management in Sri Lanka.

The CEA is the main authority with regard to enforcing the regulatory framework for waste management in the country. In the industrial estates (Export Processing Zones & Industrial Parks) managed by the Board of Investments (BOI), this function has been delegated to the BOI. Limited powers have been delegated to local authorities as well. The CEA requires all local authorities that operate sites receiving more than 100 tonnes of waste per day to obtain environmental clearance [32]. However, the CEA is unable to enforce regulations and prohibit the use of uncontrolled dumpsites throughout the country, as there are no proper alternate disposal methods.

The local authorities are responsible for waste collection and disposal within their territories. Few councils, including the Colombo Municipal council, have their own unit for waste management. Most local authorities have assigned their waste management responsibilities to the Public Health Department (PHD) under the supervision of the Public Health Inspector (PHI) [28, 32]. The PHD has many other sectors to look into, including public health and sanitation, thus waste management may not receive adequate attention.

The enforcement of law pertaining to waste management is weak and given a low priority [32]. According to the legal system, littering and unauthorized waste disposal outside designated areas are illegal. The legal system sets out very clearly the responsibility of both Government and the general public for keeping the country clean by disposing of garbage in a proper and lawful manner. Any person who dumps garbage illegally may be arrested by the police, and is liable for a fine or imprisonment [15]. Yet little attention is paid to the laws and regulations which govern waste management and disposal by the general public, authorities, and departments mandated to enforce them. The lack of enforcement is partly due to interference from people in power controlling these enterprises for their own benefit.

The National Environmental Act states that all companies and industries need to obtain licenses for dumping industrial waste that will cause pollution. However, laws and regulations are not implemented stringently by the authorities, hence, many industries continue to pollute the environment at will. The CEA considers solid waste management a task for the local authorities. According to local authorities, collection and disposal of solid waste produced by private companies is not their responsibility, rather it is the company's responsibility [28]. According to UNEP (2001), when there are no safe disposal facilities available, industries generally store hazardous waste on site without adequate management. Some industries dispose of their hazardous waste with other municipal waste, or use vacant lands to dump their waste [32]. For example, Union Carbide had a large dumpsite for battery components in a marshland in the area called Dandugama [28]. Another company, Rhino Roofing Products, was reported dumping asbestos powder and fragments in a vacant space close to a residential area causing pollution and nuisance in the area [28].

There are many legal and regulatory statutes relating to waste and hazardous waste disposal at both national and local government levels. The main legislative enactment and regulations that deal with waste management are the National Environmental Act, Provincial Councils Act, Local Government Ordinances and Hazardous Waste Regulations [32]. Moreover, Sri Lanka has signed more than 30 environmental-related international conventions and protocols including the Basel Convention on the control of transboundary movement of hazardous waste, the Montreal Protocol to control the use of ozone-depleting substances and the United Nations Framework for Convention of Climate Change [33]. None of this will be of any use or meaningful if the situation within the country remains unchanged, particularly for marginalized communities.

Many years of unsatisfactory and deteriorating services by the authorities, combined with the attitude of people, are largely responsible for the current waste management status of the country. Successive governments have been reluctant to take effective measures and enforce strict laws to resolve the problem due to the fear of losing votes. Hence, people in power attempt to transfer waste to places that will have the least effect on their popularity. Undeniably, improvements in waste management practices require firm commitment and the ability to make tough decisions by the authorities. However, this should not be at the cost of poor peoples' lives.

5.6   PRIVATIZATION OF WASTE MANAGEMENT

In this section, we analyze the private sector involvement in waste management in Sri Lanka. It aims to identify the winners and losers of privatising waste management services and considers the government's role in formulating policies and handling adverse effects of the privatisation process.

Privatisation is the transfer of state-owned resources and services to the private sector. Privatisation in developing countries often goes beyond this and refers to extending the formal systems into activities that were previously carried out by informal systems [34]. In Sri Lanka, privatisation processes were triggered after the introduction of open economic policies in 1977. However, it took another decade for privatisation to become a state policy. The slow privatisation process is largely attributed to the continued use of government-owned enterprises for the benefit of those in power, and as vehicles of employment and political support [35]. As a country in the developing region which suffered from a civil war, political unrest, high inflation, and high unemployment, such issues as who earns the benefits of privatisation and how the government uses the proceeds are important concerns [35].

Private sector participation in waste management occurs at different levels. Most local authorities conduct waste management services on their own; however, some local authorities such as Colombo and Kandy Municipal Councils have outsourced waste management to private sector firms [15]. There are a few local private companies engaged in waste management activities, however most of the companies who are in the waste business are multinational companies. Private companies such as Neptune and Polypack are engaged in large scale recycling of materials. Geocycle, a business unit of Holcim Lanka Ltd uses more advanced waste destruction processes through co-processing using cement kilns. Companies like Abans and Burns Environmental Ltd have signed waste contracts with local authorities and provide waste collection and disposal services for the local authorities. At present, private sector involvement is increasing in every aspect of waste management and there are success stories. However, this section looks at the negative consequences of privatisation when privatising of public services is carried out without proper planning.

One of the main arguments put forward by the local authorities for the privatising of waste management is that private companies provide a better quality service. However, as mentioned earlier in this chapter, waste management in developing countries is characterized by the presence of an informal sector. Granting formal contracts to waste collection services displace many waste pickers who were already performing these tasks informally. Local authorities consider the informal sector as a nuisance. Hence, their service is never considered vital, nor their opinions. Privatisation of municipal waste services negatively affects the more vulnerable group of waste pickers [36]. Some private waste management companies hire informal waste pickers. This is not common as most private companies have their own set of requirements with respect to new recruitments. However, waste pickers employed by the private companies seem to earn slightly more compared to local authority workers and street waste pickers.

Van Zon (2005) documents that private companies seem to perform their duties somewhat more efficiently than the public collection system [28]. According to a survey carried out in one of the urbanized areas in the Western province, the local residents confirmed this and stated that private company cleaners perform their duties better than the local authority workers. However, it is interesting to note that some of these private companies have no proper final disposal method and usually dump the collected waste in open dumpsites. Many residents are not aware of these malpractices.

Waste management raises two entirely different interests for both private and government sectors. For the private sector, the main interest lies with the profit the company will gain from providing the service. For the government the main concern would be to use the limited funds allocated for waste management cautiously in order to keep the city clean. Therefore, not surprisingly these two interests often clash when a government institution is obtaining the services of a private organization for waste management. We have elaborated on this using a case study of a waste management contract between a Municipal Council and a private waste management firm.

5.6.1   COLOMBO'S GARBAGE MOUNTAIN—THE BLOMENDAHL WASTE DUMP

A disagreement between the Colombo Municipal Council (CMC) and Burns Environmental Ltd. over the clearing of garbage at Blomendhal Road, the main garbage dumping site in Colombo, resulted in the residents of the area undergoing many difficulties in 2005 [37]. The municipal council could not take any action initially as the ownership of the property belonged to the private company, Burns Environmental Ltd. This was a huge open garbage mountain and during the rainy season, water from the garbage dump was flowing into the houses of residents close by who were mostly low-income residents. The flooding resulted in the spread of many diseases in the area such as diarrhoea [37].

According to the contract between the municipal council and the private company, the company was required to convert garbage into compost. Nevertheless, since the contract came into effect in 2003, the private company had not operated its compost plant at Sedawatte, one of the biggest compost plants in the region at that time. The Colombo Municipal council had paid 34–36 Million Sri Lankan Rupees monthly for its privatised garbage collection and disposal program. The payment was made for the company's services for almost two years from May 2003 to April 2005, despite the fact that the company had not carried out its contractual obligations [38]. According to a renowned environmental activist, Dr. Ajantha Perera, much of this money could have been saved had the local authorities handled the garbage collection and disposal themselves [39]. She has further stated, "the crux of the problem lies in the fact that local authorities with no knowledge of the subject are blindly following the advice of "foreign" donors and privatising garbage collection and disposal operations. The problem, common to South and East Asian countries is rooted in World Bank advisories. Privatisation is welcomed by decision making officers in air-conditioned rooms waiting for an opportunity to line their pockets" [39].

In 2007, there was a fire in the dump site which destroyed 43 houses [37]. There were around 10,000 families living in the area surrounding the dump, with approximately 1,500 families living very close to the dump, who were reluctant to leave their homes. Disaster Relief Services Centre took steps to treat residents of the area for asthma and bronchial infections caused by the fumes [37]. Again, in 2009, about 40 residents were rendered homeless when part of this dump collapsed on the shantytown on one side of the dumpsite. Lack of knowledge, disregard of local expertise, inefficiency, and private interests of officials had made many residents suffer while the Municipal councils and private companies accuse each other in the courts. This is just one example which highlights the negative implications of privatising public services without proper planning. Lack of proper planning and monitoring in the initial stages of the project has led to dishonor contractual obligations and disregard environmental regulations by all parties involved. While the parties in power argue over financial issues, waste will continue to become a mounting problem creating more chaos.

5.7   INFORMAL WASTE PICKERS IN SRI LANKA

This section presents a synopsis of the surveys, interviews, and observations carried out by the author on the informal waste pickers in Sri Lanka. For the purposes of this study, the term "informal" refers to those who generally make a living from waste but are not formally in charge of providing the service such as having employed by a municipality or having a contract with a private waste management company.

Of all the people dealing with the waste economy in Sri Lanka, the most invisible and neglected group is the informal waste collectors [19]. We see them on the roads almost every day, but little is known about their lives. Who they are, where they live, what they do or how they contribute to the waste management system in the country are questions that do not seem to concern anyone in the society. The words "scavenger" or "waste picker" draws a picture in the minds of many of an unclean individual, or a group of people, who rummage through waste piles seemingly doing nothing that could possibly be important to the society. They play an important role in waste management [40] although their service is unrecognized and underestimated [41, 42]. This section looks at the lives of this vulnerable group of people and the factors that have made them choose one of the hardest occupations in the world.

Due to the unorganized and scattered method of waste picking activities, and the lack of any formal census or literature, the exact number of waste pickers in Sri Lanka is not known. However, a rough estimation is that there are 10,000 waste pickers operating throughout the country [43]. Clearly, many variations exist as to gender, age, and the areas where they operate, but it appears that the majority of waste pickers are men aged between 18 and 35 [27]. There are very few women and child waste pickers on the streets; they are more usually working in dumpsites.

Informal waste pickers in Sri Lanka can be broadly categorized into several different groups based on how they operate. It is worth noting that there are different levels within each group. In this section, we examined the dimensions of the informal waste pickers in general as well as with specific reference to the most vulnerable groups of waste pickers in Sri Lanka.

- There are those who are well organized and carry out door-to-door collection of recyclable items. They are usually welcomed at houses and considered as carrying out a fairly dignified job.

- There is another category, those who roam the streets rummaging through waste piles searching for recyclables.

- The third group consists of those who pick waste from designated places such as municipal dumpsites, industrial dumpsites, and/or dumpsites within the Export Processing Zones.

- In addition, there are community-based organizations (CBOs) who are engaged in waste collection and sorting activities.

Waste picking is not an occupation of choice, rather it becomes an alternative when other job opportunities are lacking [41]. Sri Lanka is a country with a literacy rate of 97% with a free education system. Yet, most waste collectors have not completed basic educational qualifications or are drop-outs from schools due to extreme poverty in their families. The low literacy level is one reason that the waste pickers have chosen to scavenge through waste for their income, as most of them are unable to obtain "proper" jobs, which require some educational qualification. Many other factors that encourage them to take up waste picking as the only survival mechanism rests in the very nature of the waste collecting activity itself, which requires no skill, no capital, and no contacts to find the job. The latter is the main difference between formal and informal waste collectors in Sri Lanka, where many formal waste collectors who work for the local councils are political appointees.

The informal waste pickers provide an essential service to Sri Lanka's waste management system. However, no one seems to appreciate or see the importance of what they do. The adoption of waste picking as an occupation usually implies a downward social movement. The lack of social respect and dignity given to waste collectors completely marginalizes and erodes them from society, where ultimately they are labeled as outcasts or social pests. They are looked upon by the society as dirty and even treated as drug addicts and thieves. Although they have not resorted to begging for their daily survival, society often treats waste pickers far worse than beggars.

Among the social costs of waste collection, no less important is the cost of degraded living conditions. Most informal waste pickers are the poorest of the poor in society and live in close proximity to dumpsites, or in slum settlements. Some wander the streets during daytime to collect wastes and sleep on the street pavements. This is a very common scene around Colombo and other main urban cities in the country. The deprived living environments and lack of public services such as clean water and proper sanitation facilities create unhealthy living conditions and promotes poor health among the informal waste pickers. They suffer from intestinal and respiratory infections, skin disorders, and eye infections [8]. Most vulnerable are the children of these informal waste communities who are left alone due to the lack of support provided by the traditional family structure within the Sri Lankan society. Often this affects their education, health, nutrition, and social values and encourages them to follow in the footsteps of their parents as they grow.

When reading this section, one might question why these people choose waste collection as their livelihood if it means that their economic and living conditions do not improve. For many of them, self-employment seems to be the only option available. Waste picking is one of the very few occupations that does not require any pre requisites such as capital, special knowledge, skills, or experience. It also provides complete flexibility as to where, how, and what time of the day to work. Waste pickers do not have to abide by any rules and regulations, or follow any safety measures such as wearing personal protective clothing or obtaining medical certificates. These factors have led the society to judge waste pickers as a group of lethargic and uneducated people. Many have categorized waste pickers as a group of people who try to make money for their illegal activities. What most people do not know is that many waste pickers resort to scavenging after trying out other forms of employment or as an alternative occupation to supplement their livings.

Informal waste pickers undergo many hardships due to their livelihood more than anyone in the society. They already suffer from poverty, which has a bearing on their level of education, nutrition, and access to health services. In addition, they are exposed to many negative situations arising as a result of their daily income activities. Waste collection is a very labor-intensive job. Waste pickers usually have to search for recyclables from heaps of refuse in varying conditions of decay. Whatever the type of waste or the location, waste collection is carried out using bare hands, which makes the waste pickers unclean, and exposes them to a variety of health issues. During a survey conducted in a waste collection community it was observed that many waste collectors suffered from back pain, which is attributed to the constant bending motion required to search for waste. Carrying heavy loads of materials are associated with muscular and skeletal problems. Other major health problems reported were coughs and headaches. Some cited having sore muscles, itchy skin, and rashes while many others had suffered cuts to their hands and feet due to sharp objects. Many of them did not have access to first aid facilities in case of an emergency and were accustomed to covering their wounds with pieces of clothes.

Waste pickers have the most deprived working conditions in Sri Lanka. Often, waste dump-sites are contaminated with faecal matter, hazardous waste, and waste from hospitals. Without the protection of gloves, shoes, or masks, waste pickers handle garbage that includes human wastes, sharp metal and broken glass pieces, needles, clinical and toxic wastes. Therefore, it is not surprising that they suffer from skin diseases, respiratory disorders, and cuts from sharp objects, exposure to fumes and poisoning from hazardous waste during collection. They are exposed to illnesses and infections combined with inadequate hygiene facilities and limited access to medical care. Yet, waste collection is the only livelihood of many waste pickers and they tend to neglect their health over their work.

In addition to the inherent negative aspects of their work, waste pickers face constant harassment from municipal staff workers and from the people in the area they are operating. Municipal workers, particularly the street sweepers shout and hit their informal counterparts for picking up waste. This is mainly due to conflicts of interest over recyclables. Moreover, informal waste pickers are often held responsible for any theft taking place in the area and are typically the first suspects of the police. Most of the waste pickers travel on foot through their collection routes, as they are not welcome on public transport. In the industrial and waste collection sites, waste pickers usually rush toward the waste trucks to begin sorting through the waste that has been unloaded. This sometimes causes collisions with each other and makes it difficult to avoid accidents. Harassment is something most female waste pickers experience among the negative aspects of their work. This comes in the form of verbal and physical harassment by male collectors and sometimes competition from other waste collectors over waste materials.

Informal waste collection raises many questions. What happens after the collection process? To whom do the waste collectors sell the recyclables? How much are they being paid? How much do the buyers earn from the recyclables? We have addressed some of these questions below.

We tend to think that the waste pickers are independent workers. However, a closer look into their lives and their daily routine will prove that they are in fact dependent on others. What they need to collect, how much they are being paid, and price fluctuations of the recyclable items are all determined by external factors, thus making waste pickers far from being autonomous. Waste pickers are the main suppliers of materials to the recycling industries in Sri Lanka [19]. Most waste collectors cannot reach these industries to sell their recyclables. Hence, waste collectors have to sell their recyclables to middlemen or dealers who maintain close networks with industries. Some waste pickers sell their recyclables directly to small recycling plants, but these numbers are few.

Dealers try to maximize their profit margins by keeping payments to the waste pickers as low as possible. Surprisingly, a study carried out in a waste picking dumpsite in Sri Lanka highlighted that the links between waste pickers and the dealers are strong although dealers pay little for the waste pickers' daily collections. This is due to the power that most dealers exercise within particular areas. The internal politics of the waste economy and financial strengths of the dealers play an important role in the whole process. Dealers assure financial security to the waste pickers, particularly money-lending in times of urgency, which necessitates loyalty to the dealer. This disparity in power and wealth has made the informal waste collectors the most exploited sector in the waste economy with no bargaining power.

It is significant in the context of this discussion to note that income from waste collection is low and the working conditions are undesirable. These marginalized sectors are threatened in many ways. The public authorities pay little respect to waste pickers for the service they render in keeping the cities clean. They often take measures that have negative effects on the work of informal waste collectors as the informal waste pickers are considered as a nuisance to the environment and a threat to the image of the city. In addition to the social stigma they encounter every day, their daily collections are reducing greatly due to privatization of waste activities. Many industries, and even households, are now interested in the monetary value of recyclables and are willing to sell them to companies who buy or pay a better price for the recyclables. Hence, a new competition has arisen between informal sectors and the private companies over the same resource. Waste collectors will have little chance of survival unless they find a way of recycling the materials themselves or adding value to the recyclables in some way. There are organizations assisting informal groups to improve their livelihoods. The next chapter of this book presents a case study on the potentiality of implementing a poverty reducing solution for the informal waste sector in Sri Lanka.

REFERENCES

[1] The World Bank, *What a Waste : Solid Waste Management in Asia*, 1999, Urban Development Sector Unit, East Asia and Pacific Region. 87, 89, 90

[2] Suzuki, D.T., A. Mason, and A. McConnell, *The Sacred Balance: Rediscovering Our Place in Nature* 2007, Greystone Books. 87

[3] Population Reference Bureau, *2012 World Population Data Sheet*, 2012, Washington, D.C., USA. 88

[4] Kriner, S., Hundreds May Be Dead in Philippines After Rain-Triggered Landfill Collapse, in *Relief Web* 2000, DisasterRelief.org (The American National Red Cross). 88

[5] Petley, D., Garbage dump landslides, in *The Landslide blog* 2008, American Geophysical Union. 88

[6] Zurbrügg, C. Urban Solid Waste Management in Low-Income Countries of Asia - How to Cope with the Garbage Crisis, in *Scientific Committee on Problems of the Environment (SCOPE), Urban Solid Waste Management Review Session*, 2003. Durban, South Africa. 89

[7] Asian Institute of Tecnology, Municipal Solid Waste Management in Asia 2004, Thailand: Environmental Engineering and Management, School of Environment, Resources and Development, Asian Institute of Technology. 89, 90, 93, 96

[8] Asian Productivity Organization, *Solid Waste Management: Issues and Challenges in Asia*, M. Environmental Management Centre, India, 2007, Asian Productivity Organization: Tokyo. 89, 90, 93, 96, 102

[9] Medina, M., *Waste Picker Cooperatives in Developing Countries*, in *Wiego/Cornell/SEWA Conference on Membership-based Organisations of the Poor* 2005: Ahmedabad, India. 89, 90, 91

[10] Agamuthu, P., et al. Sustainable Waste Management - Asian Perspectives, in *The International Conference on Sustainable Solid Waste Management,*. 2007. Chennai, India. 89, 91

[11] Visvanathan, C. and U. Glawe, Domestic Solid Waste Management in South Asian Countries – A Comparative Analysis, in *3 R South Asia Expert Workshop* 2006: Kathmandu, Nepal. 90

[12] Cointreau, S., Solid Waste Collection Systems in Developing countries, in *Seminar on Solid Waste Primers and Lessons Learned from World Bank Projects*, March 7, 2005, The World Bank. 90

[13] Visvanathan, C. and J. Tränkler, Municipal Solid Waste Management in Asia - A Comparative Analysis, in *Sustainable Landfill Management*, 2003, Chennai, India. 90, 93

[14] Jayasinghe, R.A., N.J.G.J. Bandara, and W.A.S.S. Dissanayake, Current Status of Plastic Packaging Materials in Sri Lanka, in *International Forestry and Environment Symposium* 2010, Department of Forestry and Environmental Science, University of Sri Jayewardenepura, Sri Lanka. 90

[15] Environmental Foundation LTD. *Climbing out of the Grabage Dump - Managing colombo's solid waste problem*. 2007 Available from: http://www.efl.lk/publication/. 90, 93, 94, 96, 98, 99

[16] Wilson, D.C., C. Velis, and C. Cheeseman, Role of informal sector recycling in waste management in developing countries, *Habitat International*, 2006. **30**(4): pp. 797–808. DOI: 10.1016/j.habitatint.2005.09.005 91

[17] Sirimane, S., Sri Lanka the most sought after destination, in *Sunday Observer* 2012, The Associated Newspapers of Ceylon Ltd., Colombo. 92

[18] World Bank. *Data by Country - Sri Lanka*. 2011 Available from: http://data.worldbank.org/country/sri-lanka. 92

[19] Perera, A. Our Responsibility to Manage Garbage must Come Long before the Landfill Site, in *Proceedings of the International Conference on Sustainable Solid Waste Management*, 5 - 7 September, 2007, Chennai, India. 92, 101, 104

[20] Ministry of Environment and Natural Resources, *National Policy on Solid Waste Management*, 2007, Ministry of Environment and Natural Resources: Colombo. 93

[21] Bandara, N.J.G.J. Municipal Solid Waste Management - The Sri Lankan case, in *Developments in Forestry and Environment Management in Sri Lanka*. 2008. Department of Forestry and Environmental Sciences, University of Sri Jayewardenepura, Sri Lanka. 93, 94, 95

[22] Mannapperuma, N. and B.F.A. Basnayake, Institutional and Regulatory framework for Waste Management in the Western Province of Sri Lanka, in *Sustainable Solid Waste Management*, 2007, Chennai, India. 93

[23] Department of Census and Statistics, *Census of Population and Housing 2001: Population, Intercensal growth and average annual rate of growth by district, 1981 and 2001*, 2001: Colombo. 93

[24] ANZDEC Limited, *Democratic Socialist Republic of Sri Lanka: Delivering Natural Resource and Environmental Management Services Sector Project*, 2005: Colombo. 93

[25] Mudalige, D., SL's first ever sanitary landfill site comes up in Dompe, in *Daily News* 2012, The Associated Newspapers of Ceylon Ltd., Colombo. 93

[26] Clapp, J., The distancing of waste: Overconsumption in a global economy, in *Confronting Consumption*, T. Princen, M. Maniates, and K. Conca, Editors. 2002, MIT Press: Cambridge. p. 155–176. 94

[27] Jayasinghe R.A., *Informal Waste collectors' survey*, 2011, Western Province, Sri Lanka. 95, 101

[28] Van Zon, L. and N. Siriwardena, *Garbage in Sri Lanka - An Overview of Solid Waste Management in the Ja-Ela Area*, 2000, Integrated Resources Management Programme in Wetlands (IRMP): Colombo. 96, 97, 98, 99

[29] Abeysuriya, T.D., *National Reports; Sri Lanka*, in *Solid Waste Management: Issues and Challenges in Asia*, E.M.C. India, Editor 2007, Asian Productivity Organization: Tokyo. 96

[30] Perera, K.L.S. An Overview Of The Issue Of Solid Waste Management In Sri Lanka, in *Proceedings of the Third International Conference on Environment and Health*, Chennai, India, 15 -17 December, 2003. Faculty of Environmental Studies, York University and Department of Geography, University of Madras. 96

[31] Colombopage, *Dengue cases declining in Sri Lanka*. September 14, 2011. 96

[32] UNEP, *Sri Lanka: State of the Environment*, 2001, United Nations Environmental Programme, Regional Resource Centre for Asia and Pacific. 96, 97, 98

[33] GAJMA & CO Chartered Accountants, *Doing Business in...MGI International Tax and Business Guide: Sri Lanka*, 2002. 98

[34] Samson, M., Confronting and Engaging Privatisation, in *Refusing to be Cast Aside: Waste Pickers Organising Around the World*, M. Samson, Editor 2009, Women in Informal Employment: Globalizing and Organizing (WIEGO): Cambridge, MA, USA. p. 75–82. 99

[35] Malathy, K.-J. and P.P.A. Wasantha Athukorala, Assessing Privatization in Sri Lanka: Distribution and Governance, in *Reality Check: The Distributional Impact of Privatization in Developing Countries*, J.N.a.N. Birdsall, Editor 2005, Center for Global Development, Washington D.C. 99

[36] Samson, M., *Dumping on women: Gender and privatisation of waste management*, 2003, Municipal Services Project (MSP) and the South African Municipal Workers' Union (Samwu), South Africa. 99

[37] Mendis, R., Residents suffer in silence while garbage piles up, in *The Sunday Leader Online* 2005, Leader Publications (Pvt) Ltd., Colombo. 100, 101

[38] Fuard, A., Worms of a dirt deal : CMC pays millions for a company to turn garbage into compost - but it never took place, in *The Sunday Times* 2005, Wijeya Newspapers Ltd., Colombo. 100

[39] Fernando, V. and R. Hassan, Colombo's stinking nightmare, in *Sunday Observer Online* 2005, The Associated Newspapers of Ceylon Ltd., Colombo. 100

[40] Moreno-Sánchez, R.D.P. and J.H. Maldonado, Surviving from garbage: the role of informal waste-pickers in a dynamic model of solid-waste management in developing countries. *Environment and Development Economics*, 2006. **11**: p. 371–391. DOI: 10.1017/S1355770X06002853 101

[41] Atienza, V., Sound Strategies to Improve the Condition of the Informal Sector in Waste Management, in *Eria Research Project 2009; 3R Policies for Southeast and East Asia*, M. Kojima, Editor March 2010. p. 102–142. 101, 102

[42] Rouse, J., Embracing not Displacing: Involving the informal sector in improved solid waste management, in *CWG – WASH Workshop Kolkata, India*, 2006. 101

[43] Perera, A., Personal communication, *Informal waste pickers in Sri Lanka*, February 3, 2012. 101

CHAPTER 6

Assessing the Feasibility of Waste for Life in the Western Province of Sri Lanka

Toni Alyce Smythe

6.1 INTRODUCTION

In the previous chapter, we looked at waste management in the global South, and the importance of looking at this problem through a lens of social and environmental justice. The impacts of poor waste management are not felt evenly by all of society in these countries; rather, some groups, in particular waste pickers and the poor, suffer disproportionately from waste problems. Due to this inequality, even more so than in the North, solutions to waste management problems in the global South must not only address the technical but also the social side; we must ask ourselves "who pays and who benefits" and work with the former of these parties to improve their lives.

In this chapter, we will look at one example of the kinds of approaches which are being developed as socially and environmentally just responses to waste management, and assess the feasibility of such a project in Sri Lanka. Waste for Life (WfL) is an international network of engineers, scientists, academics, architects, designers, and cooperatives working to develop poverty-reducing solutions to environmental problems [1]. The goals of WfL are to:

1. promote self-sufficiency and economic security in vulnerable communities which rely on waste as a source of income; and

2. reduce the environmental impact of plastic waste [1].

Engineers play a key role in WfL, in the design of low-cost technologies to convert waste to composite products, providing an improved source of income for workers in the informal waste sector [2]. In Maseru, Lesotho, and Buenos Aires, Argentina, WfL has worked with co-operatives to develop and implement compression moulds or hotpresses to produce natural fiber composites (NFCs) from locally available waste fiber and plastic [2]. The production of NFCs can combine the excellent mechanical properties of fibers with the bonding and physical properties of plastics [3], and as a result NFCs can have advantages over conventional plastic materials, as they are light, inexpensive and more recyclable [4, 5]. They are also stronger than recycled plastics alone. The NFC products of WfL Lesotho and Buenos Aires have included insulating roof tiles, wallets, and waste bins. By working with WfL, and producing and selling useful products directly to the community, informal waste collectors can circumvent the middlemen and improve their livelihoods and working conditions.

WfL is considering implementing a similar project in the Western Province of Sri Lanka, where less than half of all waste produced is collected by the Local Authorities [6]. Prior to any detailed planning we were commissioned to assess the feasibility of the project, the crucial first step to ensuring that WfL Sri Lanka would be viable and would not harm those it is intended to benefit. This would be the first formal feasibility study conducted into WfL Sri Lanka. Feasibility studies or assessments are widely used by NGOs and government bodies to assess development projects and make an informed decision about whether a project should be implemented. This chapter comprises a preliminary socio-economic, technical, and environmental feasibility of WfL Sri Lanka.

Figure 6.1: Kingston Hot press (Photo credit: Eric Feinblatt, *Waste for Life.*)

Figure 6.2: Products designed by students for WfL (Photo credit: Eric Feinblatt, *Waste for Life*.)

6.2 SOCIO-ECONOMIC FEASIBILITY

To assess the socio-economic feasibility of WfL Sri Lanka, we conducted a Stakeholder Analysis, a Health and Safety Risk Assessment, and an identification and critical assessment of some potential sources of funding for the project.

6.2.1 STAKEHOLDER ANALYSIS

"Community development projects should be driven by the present needs and problems of the community targeted, not by an abstract or universal conception of basic human need"(p. 288).
[7]

It is generally assumed by organizations implementing technological or economic developments that they are improving the lives of individuals in a community [8]. However, the results of the project must be considered "improvement" not only by the implementing agencies but first and foremost by the intended beneficiaries. Therefore, to ensure that a project such as WfL Sri Lanka is successful, it is vital for the implementing agency to work *with* the intended beneficiaries in both planning and implementation. To allow WfL to enter into this kind of collaboration and co-creation on the ground in Sri Lanka, and give a preliminary assessment of how each may be impacted, we conducted a Stakeholder Analysis. This involves mapping the stakeholders in a project, and identifying their interests and the potential impact of the project upon their interests [9].

The steps used for the Stakeholder Analysis were based upon guidance published by Department for International Development, UK [9]. These are:

1. identification of all potential stakeholders;

2. identification of stakeholders' interests in regards to the project and assessment of the likely impact of the project on their interests;

3. identification of the influence and importance of stakeholders;

4. identification of assumptions made about the role of stakeholders in the project; and

5. identification of appropriate types of stakeholder participation (inform, consult, partner, and control) and stages in the project (identification, planning, implementation, and monitoring and evaluation).

Step 1: Identification of all potential stakeholders
For identification of potential stakeholders we conducted a profiling exercise. The following stakeholders in WfL Sri Lanka were identified based on a review of the available literature:

- informal waste collectors;

- waste buyers;

- plastic recycling companies;

- community-based organizations working with waste (CBOs);

- households providing waste;

- commercial establishments providing waste;

- local authorities;

- local Authority (LA) employed waste collectors;

- private waste collection companies;

- Waste Management Authority of the Western Province;

- Ministry of Environment;

- Central Environment Authority;

- Ministry of Health;

- Educational Institutions;

- WfL; and

- other NGOs.

Informal waste collectors in Sri Lanka gather recyclable and reusable waste from households (door-to-door waste collectors) or directly from roadsides and dumpsites (waste pickers). These people then sell the waste to waste buyers (middlemen) who in turn sell them to recyclers, either in Sri Lanka or overseas. Informal waste collectors in Sri Lanka make low profits, but in many cases this represents the person's full income. Formal waste collectors or "town cleaners," employed by the Local Authorities (LAs), collect waste from bags or bins, or more commonly from the roadside. These workers earn very little, and some formal waste collectors informally separate recyclables to sell to waste buyers in order to supplement their income [10].

We considered households as a potential source of plastic waste for the project. A survey conducted by Gunarathna [11] suggested that households will supply door-to-door collectors with plastic waste if plastics are available, however the coverage of door-to-door collectors in the Western Province is very low. We also considered commercial establishments as potential sources of plastics. The results of a questionnaire we gave to ten supermarkets in Colombo District showed that supermarkets dispose of plastic waste through LAs or waste buyers, or a combination of both. Almost half of the respondents said that they were willing to provide plastic waste free of charge for a new waste-based composite project.

The work of WfL provides educational institutions, such as universities, opportunities for cutting-edge, socially relevant research. Number of students in Canada, Argentina, the U.S., Italy, and Australia [2] have already been involved in projects focused on WfL. The University of Sri Jayawardenapura and the University of Moratuwa in Sri Lanka, together with the University of Western Australia, are expected to be highly involved in WfL Sri Lanka. Various NGOs with a dual focus on waste management and poverty alleviation exist in Sri Lanka and within the Western Province. Hence, a WfL project in the Western Province may also involve Sri Lankan NGOs as partner organizations or for the exchange of knowledge and experience.

In traditional community development terms, WfL is the "implementing agency" of the project. However, as in its previous projects WfL intends to take on a more participatory role whereby any products, processes and systems are co-created with the intended beneficiaries, partner NGOs and universities. While the organization's previous experiences in Lesotho and Argentina give WfL a unique understanding and experience of how informal waste groups can transition to produce NFCs, WfL will need to be informed by and adapted to the Sri Lankan context by utilizing local knowledge.

Step 2: Identification of stakeholders' interests in regards to the project and assessment of the likely impact of the project on their interests

In Step 2, we identified the interests of each stakeholder based upon the profiling above. The relative priority of stakeholders' interests and potential impact of the project on these interests were evaluated based upon WfL's vision for the project and the outcomes of previous WfL projects (1 being the highest priority and 5 being the least). In the table, a plus sign means positive impact, a minus sign means negative impact, and a blank means no impact on the stakeholder.

We then used a stakeholder matrix to complete Step 3, the identification of each stakeholder's relative influence (to what extent they may affect the project) and importance (the degree to which they are affected by the project).

Step 3 : Identification of the influence and importance of stakeholders

Due to the very early stage of planning for WfL Sri Lanka, we made various assumptions about the role of stakeholders in the project (which may change as the project becomes better defined). We assumed that WfL Sri Lanka would initially involve only existing CBOs, as was the case with cooperatives in Buenos Aires and Lesotho, as CBOs are already organized and may be ready to progress from collecting and sorting waste, or the production of plastic pellets, to manufacture of NFC products. We hope that, in the long-term, individual waste pickers and door-to-door waste collectors may become involved in WfL Sri Lanka through inclusion in the group working with WfL.

As a "worst-case" scenario, we assumed that LA-employed waste collectors would be negatively affected by the project given that it may result in a decreased amount of recyclable waste available for them to collect and sell as a side-income. WfL Sri Lanka ideally would target areas which are not already covered by formal waste collection and types of waste which are not currently recycled. Similarly, we assumed that the impact upon middlemen and plastic recycling companies would be negative, however depending on how waste plastics are sourced, these stakeholders may not be affected at all. We assumed that WfL Sri Lanka would have no effect on private waste collection companies. While any increase in waste collection via informal activities could result in a decrease in the potential "market" for their services, the low coverage of waste collection services in the Western Province means that this effect would not be felt. We also assumed that WfL would work with NGOs, national, and provincial government agencies, and LAs in Sri Lanka, and (provided the project is a success) this would reflect well upon these bodies and assist them (albeit slightly) to achieve their goals.

A key outcome of our Stakeholder Analysis for WfL Sri Lanka was a "map" of stakeholder participation in the project for key stakeholders, those with a high importance or influence (Step 3), produced based on guidance from DFID [9]. As is consistent with the values of WfL, we placed a stronger emphasis on consultation and partnership with stakeholders rather than informing or controlling them. We considered it appropriate that most of the key stakeholders would take on either consultation or partnership roles throughout the project, with the exception of national government bodies, provincial government bodies, and LAs. These stakeholders have high influence but low

Table 6.1: Potential impacts upon the interests of stakeholders of WfL Sri Lanka

Stakeholder	Interests	Potential project impact	Relative priority of interests
Individual waste collectors	Income Working conditions		1
Waste buyers	Profits	-	5
Plastic recycling companies	Profits	-	5
Community Based Organizations (CBOs)	Income Status in community	+ +	1
Households	Cleanliness of neighbourhood	+	2
Commercial establishments	Profits Public image	+ +	5
LAs	Cleanliness of streets	+	4
LA waste collectors	Side-income from recyclables	-	2
Private waste collection companies	Profits		5
Waste Management Authority of the Western Province	Less waste going to landfill and dumped	+	4
Ministry of Environment	Decrease in pollution caused by poor waste disposal Public image	+ +	4
Central Environment Authority	Decrease in pollution caused by poor waste disposal Public image	+ +	4
Ministry of Health	Decrease in disease caused by plastic waste	+	4
Educational institutions	Public image Institutional learning	+ +	3
WfL	Institutional learning	+	3
Other NGOs	Institutional learning through example or collaboration	+	3

Table 6.2: Stakeholder matrix showing the relative influence and importance of stakeholders of WfL Sri Lanka

	Least influence	→	Most influence
Most affected	CBOs Other NGOs Informal waste pickers		
↑	LA-employed waste collectors Door-to-door waste collectors	Commercial establishments Households	WfL Educational institutions
Least affected	Middlemen Plastic recycling companies Private waste collection companies		National government bodies Provincial government bodies LAs

importance, and could simply be kept informed at the identification (identifying and clarifying the problems the project aims to address) and monitoring and evaluation stages, and consulted in the planning and implementation stages.

We considered it essential that those ultimately affected by the project, whose importance is high but whose influence may be low (CBOs and NGOs), should work in partnership with WfL through the planning and implementation stages. Stakeholders with high influence and importance to the project (WfL and educational institutions) are considered as partners throughout the project. These roles can be seen below in the summary matrix of recommended stakeholder participation (Table 6.3).

6.2.2   HEALTH AND SAFETY RISK ASSESSMENT

To ensure that WfL Sri Lanka does not pose any undue risk to those who it is intended to benefit, we conducted a Risk Assessment to identify and evaluate potential health and safety risks associated with working with the hotpress and producing NFCs. We used a simplified procedure adapted from

Table 6.3: Summary of recommended stakeholder participation in WfL Sri Lanka

Summary of recommended stakeholder participation in WfL Sri Lanka				
Stage	**Type of participation**			
	Inform	**Consult**	**Partner**	**Control**
Identification	National government bodies, provincial government bodies, LAs	CBOs, NGOs	WfL, educational institutions	
Planning		National government bodies, provincial government bodies, LAs	WfL, CBOs, NGOs, educational institutions	
Implementation		National government bodies, provincial government bodies, LAs	WfL, CBOs, NGOs, educational institutions	
Monitoring and evaluation	National government bodies, provincial government bodies, LAs	CBOs, NGOs	WfL, educational institutions	

Hardy [12] and the UK governmental body Health and Safety Executive [13], with the following steps:

1. identification of potential hazards and risks;

2. assessment of risks on the basis of likelihood and severity; and

3. identification of mitigation measures.

Steps 1 and 2: Identification and assessment of hazards and risks

Once we had identified the hazards and risks (as discussed below), we used a risk assessment matrix such as that shown in Hardy [12] and simplified it slightly to suit WfL's needs (see Table 6.4). This matrix assigns a risk assessment value based on the likelihood and severity of risks: those risks with an assessment value of 1–3 are considered high priority, risks rated between 4 and 6 are of moderate priority, and risks assessed at values of 7–9 are low priority risks. A summary of each risk, its cause, risk assessment, and proposed mitigation measures can be seen below in Table 6.5.

Table 6.4: Risk assessment matrix adapted from Hardy (2010); red represents high priority risks, orange, moderate priority risks and yellow, low priority risks

Likelihood	Severity		
	High	Moderate	Low
High	1	3	6
Moderate	2	5	8
Low	4	7	9

Manufacturing risks: The hot press is manufactured to allow moulds up to 60 x 60 cm size to be pressed, at pressures up to 6 MPa, maintaining temperature of up to 200°C with minimal deflection of the pressing pads (less than 0.5 mm). The pressure and mould size translate into the requirement that the total force applied be up to approximately 2 MN (200 tonnes).

As the hot plates of the required press can reach extremely high temperatures (up to 200°C) during operation [14], burns to the skin (in particular the hands and forearms) is a risk associated with using the hotpress. The hot plates also take some time to cool, so the risk is present even when the hotpress has been switched off. The poor insulation of the plates means that the whole press (which is almost entirely constructed of metal) can also reach high temperatures and thus pose a burning hazard [15]. Due to the extent to which the entire press is heated, and the exposed state of many of these heated parts, we considered the likelihood of burns to the skin to be high. Given the high temperatures reached by these parts, we considered the severity of this risk to be moderate.

Another potential risk associated with use of the hotpress is injury to limbs or fingers if they are caught within the moving parts of the hotpress, such as the hot plates or jack which drives the movement of the plates. While these mechanisms operate under high pressures, their movement is extremely slow [16] and access to the areas is limited by the design of the hotpress. As a result whilst we considered this risk to be of high severity, we considered the likelihood to be low.

The experiences of WfL in Buenos Aires have shown that electric shocks are a risk, particularly when electrical components of the (highly conductive) hotpress are poorly installed or maintained. We considered the likelihood of electrical shocks from the hotpress due to poorly installed or maintained electrical components to be of moderate in the context of a developing country such as Sri Lanka. Injuries from electrical shocks depend upon the length and severity of the shock but can include burns to the skin, burns to internal tissues, and damage to the heart (potentially causing the heart to stop) [17]. If the body is wet the shock from the hotpress would likely cause death; therefore, we considered the severity of this risk to be high.

Electrical sparks are also a possible cause of fire in the workshop, as are overheating or burning of foreign material in the hot plates [15], smoking and other open flames. Provided with a source of ignition, fire has the potential to be extremely dangerous when there are large stockpiles of plastic present, as was shown when the Reciclando Suenōs cooperative warehouse in Buenos Aires burnt down in 2007 [2]. Plastic fires spread quickly, burn at very high temperatures, are difficult to extinguish [18], and their fumes are usually toxic [19]. Due to the various potential sources of ignition in the premises of a WfL project, this risk was considered to be of moderate likelihood. As it is likely that stockpiles of plastic and fiber may be present in the workshop, we considered the severity of this risk to be high.

Working environment risks: Handling waste poses a risk to workers. It can cause infection, disease, and exposure to hazardous substances; waste may be contaminated with substances such as chemical residues, pesticides [18], or bacteria. Hazardous fumes can also be produced if plastic materials are overheated [18]. Given that the cleanliness of the waste which will be handled in the project is still unknown, it was difficult to judge the likelihood and severity of the risk of infection, disease, and exposure to hazardous substances. As a conservative estimate, we considered both the likelihood and severity to be moderate.

As in nearly all workplaces, slips and trips (falls from the same level) are a risk in the premises of a WfL project, particularly if the floor is wet or cluttered, or lighting is poor. We were unable to obtain information regarding the frequency and impacts of slips and trips in Sri Lankan workplaces; however, slips and trips are a relatively common work-related injury in Australia, accounting for approximately 13% of injuries [20]. As a result, we considered the likelihood of slips and trips to be moderate, but the severity of this risk to be low.

Step 3: Mitigation measures
We anticipated that the redesign of the hotpress to suit materials available and operating conditions in Sri Lanka would provide an opportunity to mitigate some of the health and safety risk associated with using the hotpress. We expected that design modifications to the press to be the most robust mitigation measure, as the modifications would be least easily undermined, unlike for example the use of personal protective equipment (PPE). However, given that PPE is also a more easily applied mitigation measure, we considered that a combination of design modifications and the use of PPE to be the best strategy for ensuring workers' health and safety. In particular, PPE in the form of long cuff mitts such as those common in welding [16] are being successfully used by first-year engineering students operating the UWA press [15].

The summary of all risks and hazards, as well as mitigation measures identified for each risk, can be seen in Table 6.5. The most significant health and safety risks posed to those working with the waste and hotpress were burns to the skin, electrical shocks, and fire. We expected that these risks could be effectively mitigated with relative ease. The proper training of hotpress operators and the provision of good lighting for visibility will augment the mitigation measures identified.

Table 6.5: Summary of results of the health and safety risk assessment for WfL Sri Lanka

Risk	Cause	Likelihood	Severity	Risk Assesment Level	Priority	Mitigation measures
Burns to skin	High temperatures reached by hotpress	High	Moderate	3	High	Guards, improved insulation, personal protective equipment (gloves), warning signs.
Crushing of limbs or fingers	Exposed mechanical action of jack and other parts of hotpress	Low	High	4	Moderate	Guards; emergency stop; personal protective equipment (gloves, steel-capped boots).
Electrical shocks	Poorly installed and/or maintained electrical equipment	Moderate	High	2	High	Properly earthing the press.
Fire	Flammable stockpiles of plastic and fibre combined a source of ignition	Moderate	High	2	High	Maintaining clean heated surfaces, ban on smoking and open flames in the area; ensuring flammable materials are sited sufficiently far from hotpress and other electrical systems; securing premises.
Infection disease, and exposure to hazardous substances	Contact between workers and contaminated waste materials, overheating of plastics	Moderate	Moderate	5	Moderate	Personal protective equipment (gloves, face masks); treating and covering wounds; toilet and washing facilities; regular checks of equipment operating temperatures, maintaining clean heated surfaces; adequately ventilated work space.
Slips and trips	Untidy or dirty workspace	Moderate	Low	8	Low	Ensuring floor is clear of obstacles; cleaning up spills of liquid promptly.

6.2.3   IDENTIFICATION AND CRITICAL ASSESSMENT OF SOME POTENTIAL SOURCES OF FUNDING

To assess the preliminary economic feasibility of WfL Sri Lanka, and provide a starting point for WfL to further investigate funding options for the project, we identified some possible sources of funding and assessed these on the basis of their appropriateness to WfL values and aims. Funding for

WfL Buenos Aires projects has been sourced from a micro-credit organization and an international aid program [2]. Funding for WfL Lesotho was obtained via the Small Grants Fund of the Global Environment Facility [2]. This program awarded USD 50,000 to the Cooperative College to make and maintain the hotpress and provide training to the community group (Maseru Aloe) which uses the press to produce NFCs.

There are risks associated with any source of funding, unless it is donated with no strings attached. Grant funding can be excellent, but there is a trade-off between funding agencies refusing to pay partners in the North to manage projects (for good participatory principles) and partners on the ground, being inexperienced with managing, what for them may be large funds. Microfinance is an excellent source of funding for small sums, particularly to support the purchase of equipment which can be held as collateral until the loan is paid off. The loans can act as a good incentive for motivation of local partners, rather than gift equipment which may remain unused once the first break down occurs (as has been WfL's experience).

According to Jayamana [21], the Global Environment Facility Small Grants Programme is potentially the largest provider of grants to NGOs with an environmental focus in Sri Lanka. WfL Sri Lanka may be eligible for grants from this program under the Persistent Organic Pollutants focus area (which includes waste recycling) [22]. The Global Environmental Facility has been criticised as having been designed to service "a set of global institutions that themselves exist at least partly to serve certain globalised political and business interests" (p. 17) [23]; in particular, due to its ties to The World Bank, which tends to imposes "top-down" solutions to environmental problems. We considered that securing a grant from this organization would have little influence on how WfL works or the implementation of its project in Sri Lanka, and as such this funding program remains a viable opportunity for securing funding.

Microfinance is the provision of a variety of small-scale financial services, including credit, savings, and insurance, to people who cannot access funds through formal financial institutions [24]. Usually, microfinance schemes are characterized by easy access for impoverished people, group-based lending, little or no collateral requirements and "reasonable" interest rates (pg. 2) [24]. Microfinance is widely regarded as having significant potential for poverty alleviation and development, with microfinance institutions' thought to encourage entrepreneurship and economic development "at the grassroots level" (pg. 10) [25]. There are various criticisms associated with microfinance, however, key of which is that microcredit can lead to "microdebt" and leave borrowers in more difficult financial situations than before the loan [26]. Contrary to expectations, microfinance can be expensive for borrowers; the global average interest rate for microfinance is approximately 35% [27]. While microfinance can be a highly effective tool for poverty alleviation, it is not a fool-proof solution to poverty and should be used with care [26].

The microfinance funding for WfL Buenos Aires has been successful, due in large part to the careful selection of the microfinance provider. The Working World was chosen for the project as it is a not-for-profit organization and was able to provide the loan with zero interest. The Working World was also previously involved with cooperatives in the area and worked *with* the cooperatives, providing support and training. Furthermore, if the project failed, The Working World was willing to forgive the loan and simply reclaim the press (which is worth no more than the loan).

WfL raised the funds (USD 4000) and donated the money to The Working World on the proviso that the money was loaned to the cooperative producing NFCs. Thus, The Working World acts as a conduit for funding from WfL and takes on the responsibility of managing the loan. When the loan is repaid by the cooperative in Buenos Aires, the funds will be available to lend for another WfL project.

Given the success of WfL's arrangement with The Working World, we expected that a similar situation might be beneficial for a WfL project in Sri Lanka; however, the practices of the micro-finance organization involved must be consistent with WfL's values and project aims. While in Sri Lanka average loan interest rates are 17% [27] and thus much lower than the global average, these rates still be too high for the poorest of Sri Lankan society. Therefore, we considered the interest rate at which microfinance would be provided to be a key consideration in identifying potential sources. To be consistent with WfL's values we also focused on government and not-for-profit microfinance providers, and NGOs offering microfinance.

One potential microfinance provider for WfL Sri Lanka we identified was the Jana Suwaya People Development Foundation (Jana Suwaya), an NGO established in 1994 in Hambantota on the South East coast by a local Member of Parliament. While Jana Suwaya was initially focused on community development in Hambantota, the organization has diversified to microfinance and expanded to a national level, with loan schemes already implemented in Anuradhapura, Kurunegala and Kalutara districts. According to the organization, the success of its projects can be attributed to the active participation and leadership of the local community. Of particular interest to WfL is Jana Suwaya's zero-interest loans. We speculated that as part of plans to further expand the organization and develop their zero-interest microfinance scheme, Jana Suwaya might be interested in partnering with WfL. Jana Suwaya can work with WfL in a donation-loan scheme similar to the role of The Working World in the Western Province. The values of Jana Suwaya also appear to be consistent with WfL, particularly the organization's focus on community participation and zero-interest loans. However, due to the prominent political position of the Jana Suwaya's founder and sometimes chairman [28], we recommended that the organization be investigated further to ensure its political neutrality and long-term sustainability.

We considered Samurdhi Bank Societies, government institutions regulated by the Samurdhi Authority of Sri Lanka, as another potential source of microfinance for WfL Sri Lanka. These societies were established in 1996 as part of a governmental poverty alleviation initiative [29]. Loans are available to low income-earners at interest rates of between 7 and 12% [30]. Samurdhi Bank Societies also provide planning and technical support for income generation projects [30]. There are approximately 1,000 branches of the Samurdhi Bank Societies network [31] with approximately 2.95 million active microfinance clients [32]. For group loans, each member must open an individual account and deposit 500 Sri Lankan rupees, but the group may apply for loans up to 10 times of the total value of deposits [30]. Due to the low interest rates of these loans, the abundance of branches and the additional support provided to lenders, we considered Samurdhi Bank Societies to have strong potential for a source of funding for WfL Sri Lanka.

While it is mandatory for commercial banks to allocate 5% of their portfolio to microfinance, microfinance in Sri Lanka is generally provided by non-regulated institutions, such as NGOs, societies and cooperatives [32]. There are a vast number of small-scale, local unregulated microfinance institutions operating in Sri Lanka [32]. These smaller-outreach organizations have been criticized for being "administratively weak and financially dependent on donors for sustainability" and average loan sizes are fairly low (below USD 200, or approximately 22,000 Sri Lankan rupees for 94% of MFIs) (pg. 23) [32]. We considered that these types of institutions remain a funding option for WfL Sri Lanka and should be further investigated.

6.3   TECHNICAL FEASIBILITY

To assess the technical feasibility of WfL Sri Lanka, we identified and critically assessed the various potential sources of waste plastic and natural fiber for the project. The selection of fiber and matrix will be important not only for the economic viability, environmental impact and sustainability of the project, but will also heavily influence the properties of the resulting NFC products. WfL has developed several unique NFCs for previous WfL projects: in Canada, HDPE bale wrap was reinforced with flax and hemp, in Lesotho, corn, wheat, and agave were combined with LDPE plastic bags, and in Buenos Aires NFCs were created using paper, card, or fabric, and HDPE and LDPE plastics [14].

6.3.1   IDENTIFICATION AND CRITICAL ASSESSMENT OF POTENTIAL SOURCES OF WASTE PLASTIC

Plastics are widely used in Sri Lanka due to their low cost and availability; in 2010, the total annual plastic consumption in the country was estimated at 100,000 tonnes [33]. There are between 20 and 25 types in use [34]; the most common types are shown in Table 6.6.

Table 6.6: Common types of plastic used in Sri Lanka [35]	
Type	**Use**
Polyethylene: high-density polyethylene (HDPE) and low-density polyethylene (LDPE)	Widely used for packaging, plastic bags and insulation
Polythene terephthalate (PET) [36]	Used in recyclable containers such as plastic bottles
Polypropylene (PP)	Used for packaging as relatively heavy films and woven sheets or bags
Polystyrene (PS)	Used for protective packaging and insulation
Polyvinyl chloride (PVC)	Primarily used for building materials such as pipes and flooring

In most countries, there is a well-developed market for recycled PET, so we expected that PET would be already highly utilized by recyclers in the Western Province. Using PET in NFC production would also pose a technical challenge for a WfL project as this plastic tends to crystallise during heating and become very brittle unless additives are used [37]. PVC and PS are also recycled in Sri Lanka but are generally avoided by WfL as they emit toxic fumes when overheated (discussed later in this chapter).

Of the seven recyclers which responded to a questionnaire we conducted, six recycled one or both of HDPE and LDPE plastics. This suggests that HDPE and LDPE plastics may be already widely used in existing recycling facilities in the Western Province (and thus more difficult to source for a WfL project). However, plastic in the form of films is typically difficult to collect, store and clean, so we expected the demand for HDPE and LDPE *films* in the current recycling market to be very low. Plastic films also allow for greater ease of production in the WfL hotpress, as they may be simply layered with fiber without grinding or chopping the plastic. While rigid plastics can be used in the hotpress, they must first be ground using a specialised machine to break the plastic into small pieces. We focused on assessing a selection of thin flexible plastic products (made from HDPE, LDPE, or PP) common in Sri Lanka.

One of the most common plastic products produced and used in Sri Lanka is plastic bags [33]; supermarkets in Sri Lanka provide plastic shopping bags free of charge [35]. While plastic bags are often reused in households, their low quality combined with their abundance means that this practice has little effect on the quantities available [38]. In the Western Province, approximately 1270 tonnes of plastic bags enter the MSW stream annually [33, 39]. Due to the high availability of plastic bags, and their success as a material for NFCs produced with the WfL hotpress in the past, we considered plastic bags to have strong potential for use in WfL Sri Lanka.

Supermarkets in the Western Province also generate large amounts of plastic waste in the form of primary (for bulk food goods), secondary (for multiple products) and tertiary (pallet) wrappings; these can be either LDPE or HDPE films. The results of a questionnaire given to ten supermarkets in Colombo show that the amount of this type of waste generated by supermarkets can vary significantly, but may be up to 30 kg per week per supermarket. We considered plastic wraps to be an excellent source of plastic for WfL Sri Lanka as they would most likely be clean, homogenous, and generated continuously. They may also be generated in large quantities.

Other products made from plastic film may be available in significant amounts on a local scale. Most items sold in supermarkets in the Western Province are packaged in plastic materials, including milk powder, biscuits, nuts, confectionary, dried noodles, flour, grains, sugar, tea leaves, and coffee [35]. These kinds of packets are commonly made from HDPE film and, as they contain dry food goods, would be relatively easy to clean. Another possible source is plastic book covers; textbooks are provided free of charge by the Sri Lankan government to schoolchildren, but are required to be covered in thin plastic film (usually HDPE) each year before they may be taken to school. Book covers are one of the most widely used forms of plastic packaging in the Western Province [35]. Thin HDPE tablecloths are also popular in Sri Lanka and typically have a useful life of about six months. We expected that each of these products may have potential for use in WfL Sri Lanka, but further investigation into these sources is needed.

The results of the questionnaire we gave to seven recyclers in the Western Province suggested that collection from households is currently under exploited by plastic recyclers. Due to the willingness of households to supply door-to-door collectors with plastic waste and the low coverage of door-to-door collectors (discussed above as part of the Stakeholder Analysis), we considered that there was significant opportunity for collection from households without competing with door-to-door collectors. Household waste was expected to comprise a significant amount of plastic material; according to the Ministry of Environment's Database of Solid Waste [40] plastics make up 5.9% of the municipal waste collected in Sri Lanka. This figure does not take into account the significant amounts of plastics which are illegally dumped or burned.

The one major disadvantage with sourcing plastics from the municipal solid waste stream would be the high risk of contamination which could significantly affect the quality of the recycled final product [33]. Over half of the plastic recycling facilities in Sri Lanka wash their plastics [11] however this practice has a significant negative impact upon the environment in Sri Lanka (discussed later in this chapter). Collecting plastic wastes less likely to be dirtied through use and *before* they enter the mixed municipal waste stream (i.e. plastic bags from households which have "pre-sorted" their waste) would mean that waste would be relatively clean and also require little sorting [33]. Therefore, we expected that a source separation scheme similar to that organized by Arthacharya for its Galle recycling plant [41] would eliminate the necessity of extensive sorting and washing of household waste before producing NFCs. Furthermore, the involvement of the community in waste collection through source separation might not only improve the quality of waste collected but also the social standing of the collectors [42].

Household collection alone might not be able to provide sufficient quantities of plastic waste for the WfL project, as was found with the Galle Recycling Plant [41]. For the project to have a continuous supply of suitable plastics in sufficient quantities, we considered household collection would be combined with collection from one or more commercial facilities (for example the plastic wraps discussed above). We also expected that, as the community became aware of the project, some businesses or households not covered by the collection scheme might voluntarily deposit suitable plastic wastes at the site of production.

6.3.2   IDENTIFICATION AND CRITICAL ASSESSMENT OF POTENTIAL SOURCES OF NATURAL FIBER

We identified and assessed some key potential sources of waste fiber for WfL Sri Lanka based on the following considerations.

1. The materials should be underutilized and of little commercial worth.

2. Ideally the materials should be available year-round (i.e. supply should be continuous rather than seasonal) in sufficient quantities to sustain the project.

3. Due to the limited capital available for the project, the source of the materials should be relatively close to the manufacturing location.

4. For ease of production, the materials should require as few as possible additional processing steps.

Data from the Sri Lankan Department of Census and Statistics (where available) and Residue to Product Ratios (RPRs) given in previous studies or calculated using average weights, allowed us to assess the availability of various agricultural fibers by estimating the generation of each fiber in the potential source area. We considered the Western Province and surrounding districts (Puttalam, Kurunegala, Kegalle, Ratnapura, Galle, and Matara) (see Figure 6.3) due to the increase in transport costs with the increase in distance.

Figure 6.3: The source area for potential waste fibers for WfL Sri Lanka (yellow and green); the districts of the Western Province are shown in yellow.

The fibers used in previous WfL projects were sourced from dedicated crops (agave, hemp, or flax), agricultural byproducts (corn stalk or wheat straw), or post-consumer sources (cardboard, office paper, and newspaper) [14]. In Sri Lanka, there are various existing sources of natural fiber, both agricultural and post-consumer. We focused on eight common fibers with potential for use in NFCs: coir, paper products, rice husk, rice straw, sawdust, sugar cane bagasse, and textiles.

Coir is a byproduct of coconut production, and forms part of the husk of the coconut [44]. Coconut is grown extensively in Sri Lanka, particularly within the districts of Colombo, Gampaha, Kurunegala and Puttalam, a region called the "coconut triangle." Although more fruits are produced in some months than others, coconuts are produced throughout the year [45] and we estimated the annual coir production in the source area as approximately 292,000 tonnes. The fiber is strong and durable but without a specialized machine, extracting the coir is time-consuming and labor-intensive; once the husk has been removed from the fruit it is retted (generally soaked in water for up to six months), after which the fiber is extracted by beating [3, 46]. Coconut husk is also already used in the fiber industry for products such as twine and geo-textiles, used as a fuel for domestic cooking, and exported as chips [47]. Therefore, while coir is a high-quality fiber and produced year-round in large amounts, the demand for other applications meant that we expected that this fiber would have a relatively high value and be somewhat difficult to obtain. Processing of the husk would also pose a significant challenge for a WfL project.

Rice or paddy husk is the hard protective outer layer of a rice grain [48]. There is considerable transfer of rice occurs between districts [44], hence we were unable to estimate the generation of rice husk in the WfL source area using data from the Department of Census and Statistics. Rice mills are widespread in Sri Lanka; the country boasts around 7,000 mills [44]. The mills store the rice produced in the two growing seasons and operate year-round. In previous studies of NFCs, rice husk has been milled or ground before use in NFCs [49, 50, 51, 52]; this may be due to the poor chemical and physical adhesion properties of rice husks, which grinding can address [53]. Small amounts of rice husk are used for applications such as brick manufacturing [44, 54], however surplus husk is generally considered waste and is often freely available [44]. We considered rice husk as an excellent potential source of fiber for WfL Sri Lanka. It is produced year-round in large amounts and there is little demand for other applications, which means it is of low worth to producers.

Rice or paddy straw is the stalk material which remains after the rice has been harvested from the paddy [55]. Rice is grown in almost all areas of Sri Lanka, and Kurunegala is one of the major rice-producing districts [56]; we estimated that approximately 1,112,000 tonnes of rice straw is produced annually in the source area. The availability of rice straw is highly seasonal; harvesting is conducted at the close of each growing season, in February and August [44]. Like rice husk, rice straw is generally ground into particles before the production of NFCs [57], to ensure an even distribution of fiber properties, among other reasons. Currently, much of the rice straw produced in Sri Lanka is simply burnt or ploughed into the field [44, 58]. There is some small demand for rice straw in uses such as paper making and animal feed, however due to the large amounts which are generated, these uses have little effect upon its availability [44]. While we expected that rice straw could be easily sourced in sufficient volumes for WfL Sri Lanka, its seasonal availability would make this fiber unsuitable for the project as large storage areas would be required to sustain a continuous operation.

Disposing of sawdust poses a challenge for Sri Lanka's sawmills, of which there are more than 4,000 [44]. These saw mills generally operate year-round, and the main sawmilling area in the country is the city of Moratuwa, in Colombo District. A small amount of sawdust is consumed for heat generation in the manufacturing industry and for domestic purposes, and most sawdust is dumped in large heaps around the mills and is freely available [44]. In Moratuwa, sawdust is a major environmental problem as it is often dumped into rivers and the ocean. We estimated the generation of sawdust in the source area as around 51,000 tonnes. This is the lowest estimate of the fibers we looked at. Its year-round production and status as a nuisance waste indicates that it is of no commercial value and would be very readily available for the WfL project. Thus, we considered that sawdust has significant potential as a raw material for WfL Sri Lanka.

Sugar cane bagasse is the fibrous material which remains after sugar cane is crushed for the production of sugar [59]. There are three main sugar manufacturing plants in Sri Lanka; one of these, Sevanagala Sugar Industries Ltd., is located within the source area in Ratnapura District [60]. An estimated 72,500 tonnes of sugar cane bagasse is produced annually from this plant alone [60]. The only drawback is for use in NFCs it is necessary to first clean the bagasse to remove organic debris and sugar residue [61]. This may be achieved simply through washing in water [62]. The fibers are then dried and cut into uniform pieces to ensure consistent properties within the composite [62]. Most sugar factories use the bagasse they produce for an energy source, burning it to generate steam and electricity [63], with any surplus bagasse used to produce alcohol, feed animals and make compost [60]. Whilst there is a significant generation of this fiber within the source area we expected that it would not be readily available for a WfL project due to its high usage in other applications.

Paper products comprise over 6% of the Municipal Solid Waste (MSW) collected in Sri Lanka. This is the greatest contributor to MSW after short-term biodegradable waste [40]. The MSW collected within Gampaha Municipal Council in particular has a very high paper component at 18% [64]. The most common types of paper wastes in Sri Lanka are office paper, newspaper, and paper packaging. Paper packaging in particular is common in Sri Lanka; most of the products sold through small grocery shops, which low-income earners generally shop in, use paper and paper bags to package food. Paper waste requires little pre-processing as it may be incorporated into NFCs by simply cutting to size, layering and pressing; however waste paper products are already recycled in Sri Lanka through informal collection, as well as in home-based industry [65]. A better understanding of the extent that paper is already recycled in the area should be assessed to avoid any potential negative impacts on informal waste collectors and home-based paper recyclers.

Textile waste such as fabric off-cuts are a product of garment manufacturing, which is a major industry in Sri Lanka. Garments make up almost half of all exports from the country [66]. There are 12 major industrial areas in the country, most of which are situated in Western Province [67]. Biyagama and Katunayake Export Processing Zones (EPZ) in Gampaha are the largest industrial areas in Sri Lanka [68], and the waste produced in these areas is dominated by textiles which comprises approximately 50% and 60%, respectively [69]. Like paper products, textile waste may be incorporated into NFCs by simply cutting, layering and pressing. The long, woven fibers can make very strong and flexible NFCs. Waste pickers already frequent EPZ dump sites to collect waste and at Katunayake, textiles are a major component of materials collected [70]. However, while waste pickers collect large quantities of textiles, they represent only a moderate source of income when compared to other types of waste such as plastics. Given the high generation of textile waste expected at EPZs and the excellent mechanical properties of NFCs made with textiles, we considered this material to have a potential for use in a WfL project. It must be ensured that there would be no negative impact upon waste pickers who already collect textile waste for home-based industries.

As a result of our assessment, we expected that a WfL project in the Western Province would be able to source sufficient amounts of suitable waste materials to sustain the project. We considered household collection, with a source-separation scheme, combined with collection from one or more commercial facilities to be the best collection method for plastic waste. Plastic bags and supermarket wrapping are considered to have the strongest potential as a waste plastic in the WfL project due to their abundance, the difficulty in recycling these products through conventional means, and their likely cleanliness and homogeneity. Of the eight types of fiber examined, we considered rice husk and sawdust have the greatest potential for use in the project, due to the large amounts generated year-round in the source area, low demand in other applications (and therefore low commercial value), and suitability for use in NFCs with little extra processing. We also considered textile waste from EPZs to have significant potential as a source of fiber provided any negative impacts upon waste pickers were prevented.

6.4 ENVIRONMENTAL FEASIBILITY

While recycling can have significant environmental benefits, the industry is also associated with some environmental costs. The implementation of a WfL project in the Western Province could also involve a CBO which is already practicing some kind of plastic recycling. Therefore, to assess the environmental feasibility of WfL Sri Lanka we compared the environmental impacts of conventional plastic recycling in the Western Province with the WfL process of NFC production.

6.4.1 COMPARISON OF CURRENT PRACTICES IN PLASTIC RECYCLING WITH WASTE FOR LIFE PROCESS

We compared the WfL NFC production process to current plastic recycling practices in the Western Province in terms of:

1. air pollution;

2. energy use; and

3. water use and pollution.

Many plastic recyclers in Sri Lanka process PS and PVC among other types of plastics (55% and 24% of recyclers, respectively) [11]. These types of plastics are generally avoided by WfL due to their tendency to release harmful gases and fumes when overheated, including carbon dioxide (CO_2), carbon monoxide (CO), and hydrochloric acid (HCl) [71, 72, 73]. Carbon dioxide is a greenhouse gas and CO is a precursor to greenhouse gases as it increases the concentration of methane and ozone in the atmosphere, and eventually oxidises to produce CO_2 [74]. HCl has an acute toxic effect and contributes to the processes which cause photochemical smog and acid rain [75]. We expected that the WfL manufacturing process would most likely result in less air pollution than current recycling practices in Sri Lanka, as the exclusion of PS and PVC from the WfL process avoids the production of HCl, CO and CO_2 gas from these sources.

The incorporation of natural fibers into WfL's products can also act as a "sink" for atmospheric carbon dioxide due to the stored carbon within the fibers [76]. In particular, if the project made use of a fiber which is currently burnt, such as rice husk this would prevent the release of combustion fumes and greenhouse gases (including CO_2) which would otherwise occur.

When we embarked on the feasibility assessment we aimed to directly compare the energy use of plastic recyclers in the Western Province with the WfL process. Due to a lack of data we were unable to accomplish this comparison, current practices in Sri Lanka generally involve a relatively high number of processing steps, which suggested that their energy use would be similarly high. In the conventional plastic recycling process in Sri Lanka, plastic waste is processed by size reduction by manual cutting, mechanical shredding, or agglomeration (heating and rapidly cooling to harden soft plastics which are crushed to form coarse grains), washing, extrusion, and then pelletizing [33]. Furthermore, the end product of most of these operations is not in a usable form. Only about a third of plastic recyclers in the Western Province progress to manufacturing after pelletizing [11]; most simply sell the pellets to a manufacturer [33]. The results of our questionnaire for plastic recyclers in Colombo and Gampaha districts suggested that most of the recyclers which progress to manufacturing produce only intermediate products, which would require further processing and energy input before they are *useful products*. Given that the WfL process can be achieved with only a single mechanical processing step (hotpressing) and produces useful products, the WfL process is less energy-intensive than conventional recycling in the area, which uses different machinery in each process.

Water use and pollution is a major issue for conventional plastic recyclers in Sri Lanka. Large amounts of water are used for washing, rinsing, and cooling of plastics in the recycling industry in the Western Province [33], much of which is sourced from groundwater via wells (72%) or from rivers (14%) [11]. Washing is usually conducted by hand; a solution of alkaline detergent (i.e. laundry powder) or caustic soda in water is used to remove contaminants [33]. The resulting wastewater, which is generally of poor quality, is discharged without treatment and usually without reuse into nearby surface water bodies, unsealed pits in the ground, or stormwater drainage pipes which discharge to surface water bodies [33]. Previous WfL projects and testing has demonstrated that it is generally unnecessary to prepare waste plastic used in the organization's process through intensive cleaning [14]. As such we expected that the WfL process would represent a significant saving of water and reduction in pollution of soil, groundwater, and surface water when compared to existing recyclers. WfL Sri Lanka also has potential to prevent existing water pollution if the plastic and fiber source used would otherwise be dumped in waterways, for example sawdust.

While we were unable to make a direct quantitative comparison, given the differences in WfL's NFC production process and the process employed by the majority of plastic recyclers in the Western Province, we expected that a WfL project will have some environmental advantages over current plastic recycling operations in the area, through decreases in water use and pollution, air pollution and potentially, energy use. We considered that the project could provide further environmental benefits by providing a recycling avenue for fibers which are currently dumped or burned. Ideally WfL would also use plastics which are not currently recycled by conventional means, such as thin HDPE or LDPE films.

6.5   FUTURE DIRECTIONS FOR WFL SRI LANKA

WfL Sri Lanka is an excellent example of a socially just approach to waste management as WfL works with the informal sector in waste management. From the results of the preliminary feasibility assessment we conducted for WfL Sri Lanka we concluded that that the project could be considered highly feasible, warranting further, more exhaustive studies incorporating data collection on the ground.

In particular, we expected that quantitative data would be able to provide a more accurate assessment of technical and environmental feasibility. Further investigation into local small-scale, unregulated microfinance providers in the Western Province would provide the basis for a more comprehensive assessment of potential sources of funding, and personal communication with stakeholders would allow a more in-depth Stakeholder Analysis with identification and profiling of specific stakeholders (rather than broad groups).

We also considered that the final selection of materials should be an iterative process which occurs concurrently with the design of the NFC product, which is undertaken with and is primarily driven by the needs of the intended beneficiaries. As this is being done, it would help to undertake further research to determine the *local* availability of specific fibers and plastics and to characterise the properties of NFCs produced from potential fibers and plastics. This work will be undertaken as part of more detailed feasibility studies conducted at later stages throughout the project's planning and implementation to ensure the continued feasibility of WfL Sri Lanka as a socially just response to waste management.

REFERENCES

[1] *Waste for Life*. 2011. Available from: `http://wasteforlife.org/`. 109

[2] Baillie, C., et al., Needs and Feasibility: A Guide for Engineers in Community Projects—The Case of Waste for Life. Lectures on Engineers, Technology, and Society, 2010, San Rafael, California: Morgan and Claypool. DOI: 10.2200/S00249ED1V01Y201001ETS013 109, 113, 119, 121

[3] van Rijswijk, K., W.D. Brouwer, and A. Beukers, *Application of Natural Fibre Composites in the Development of Rural Societies*, 2001, Delft University of Technology: Delft. 109, 127

[4] Torres, F.G. and M.L. Cubillas, Study of the interfacial properties of natural fibre reinforced polyethylene, Polymer Testing, 2005. **24**(6): pp. 684–698. DOI: 10.1016/j.polymertesting.2005.05.004 109

132 REFERENCES

[5] Thamae, T. and C. Baillie, A life cycle assessment of wood based composites: A case study, in *Wood-Polymer Composites*, K. Oksman and M. Sain, Editors. 2008, Woodhead Publishing Limited: Cambridge. DOI: 10.1201/9781439832639 109

[6] United Nations Environment Programme (UNEP), *3.2 Waste Disposal*, in *Sri Lanka: State of the Environment 2001*, United Nations Environment Programme. pp. 42–52. 110

[7] Manzo, K., Nongovernmental Organizations and Models of Development in India, *Journal of Environment and Development*, 2000, **9**(3): pp. 284–313. DOI: 10.1177/107049650000900305 111

[8] Reviere, R., et al., eds. *Needs Assessment- A Creative and Practical Approach for Social Scientists.* 1996, Taylor and Francis: Washington, D.C. 111

[9] Department for International Development (DFID), *Guidance Note on how to do Stakeholder Analysis of Aid Projects and Programmes*, 1995, Department for International Development: London. 112, 114

[10] van Zon, L. and N. Siriwardena, *Garbage in Sri Lanka: An Overview of Solid Waste Management in the Ja-Ela Area*, 2000, Integrated Resources Management Program in Wetlands: Colombo. 113

[11] Gunarathna, G.P.N., *Analysis of Issues and Constraints Associated with Plastic Recycling Industry in Sri Lanka*, 2010, University of Sri Jayewardenepura: Nugegoda, Sri Lanka. 113, 125, 129, 130

[12] Hardy, T., *The Role of Human Factors in Safety Risk Assessment*, 2010, Great Circle Analytics LLC: Denver, Colorado. 117, 118

[13] Health and Safety Executive, *Five steps to risk assessment*, 2006, Health and Safety Executive: Bootle, England. 117

[14] Baillie, C., et al., Waste-based composites - Poverty reducing solutions to environmental problems. *Resources, Conservation and Recycling*, 2011, **55**(11): pp. 973–978. DOI: 10.1016/j.resconrec.2011.05.006 118, 123, 126, 130

[15] Tavner, A., Personal communication, *Identification and assessment of hazards and risks of using the hotpress*, September 21, 2011. 118, 119

[16] Matovic, D., Personal communication, *Identification and assessment of hazards and risks of using the hotpress*, September 1, 2011. 118, 119

[17] Comcare. *Electrical work - What are the OHS risks?* Available from: http://www.comcare.gov.au/safety__and__prevention/health_and_safety_topics/electrical_safety/electrical_work_-_what_are_the_ohs_risks. 119

[18] Health and Safety Executive, Plastics recycling, in *Plastics Processing Sheet*, No 2, 1998, Health and Safety Executive: Bootle, UK. 119

[19] Health and Safety Executive, *Designing and operating material recycling facilities (MRFs) safely*, 2009, Health and Safety Executive: Bootle, UK. 119

[20] WorkSafe. *WorkSafe operational priority statistics summary*, 2011; Available from: `http://www.commerce.wa.gov.au/worksafe/content/Services/Facts_and_figures/Priority_areas.html`. 119

[21] Jayamana, M., *The Civil Society and Environmental Movement in Sri Lanka: Sri Lanka Country Analysis*, 2008, Development Fund: Norway. 121

[22] Global Environment Facility (GEF) Small Grants Programme. *Persistent Organic Pollutants.* 2006; Available from: `http://sgp.undp.org/index.cfm?module=projectsandpage=FocalAreaandFocalAreaID=POP`. 121

[23] Young, Z. and S. Boehmer-Christiansen, Green Energy Facilitated? The Uncertain Function of the Global Environment Facility, *Energy and Environment*, 1998. **9**(1): pp. 35–59. 121

[24] Colombage, S.S., Microfinance as an Instrument for Small Enterprise Development: Opportunities and Constraints, in *The 23rd Anniversary Lecture of the Centre for Banking Studies*, 2004, Central Bank of Sri Lanka: Colombo. 121

[25] Hudak, K., Political institutions and grassroots development: the political economy of microfinance, in *Department of Political Science*, 2010, Northeastern University: Boston, Massachusetts. 121

[26] Srivastava, T., *Microfinance: A Comparative Analysis of Varying Contexts, Current Needs, and Future Prospects between Developing and Developed Countries*, 2010, Bowling Green State University. 121

[27] Kneiding, C. and R. Rosenberg, *Brief: Variations in Microcredit Interest Rates*, 2008, Consultative Group to Assist the Poor: Washington D.C. 121, 122

[28] United National Party. Sajith Premadasa's Jana Suwaya sans Govt. backing helps all, young and old. 2007; Available from: `http://www.unp.lk/portal/index.php?option=com_contentandtask=viewandid=490andItemid=2`. 122

[29] South Asia Microfinance Network (SAMN). Microfinance in Sri Lanka; Available from: `http://www.samn.eu/?q=srilanka`. 122

[30] Samurdhi Authority of Sri Lanka. *FAQs.* 2011; 2011 25 September]; Available from: `http://www.samurdhi.gov.lk/web/index.php?option=com_contentandview=articleandid=91andItemid=84andlang=en`. 122

134   REFERENCES

[31] Samurdhi Authority of Sri Lanka. *Samurdhi Bank Societies*. 2010; Available from: http://www.samurdhi.gov.lk/web/index.php?option=com_contentandview=articleandid=100andItemid=78andlang=en. 122

[32] Gomez, A., Microfinance Sector Assessment, in *Connecting Regional Economies*, 2009, USAID and AECOM International Development. 122, 123

[33] Jayasekara, P.M., Water Pollution Associated with Plastic Recycling Industry in Sri Lanka, in *Department of Forestry and Environmental Science*, 2010, University of Sri Jayewardenepura: Nugegoda, Sri Lanka. 123, 124, 125, 130

[34] Lakmali, W.A.S. and A. Dissanayake, *Study Report: Understand the Effectiveness on Thin Polythene Regulation*, 2008, Central Environment Authority, Government of Sri Lanka. 123

[35] Bandara, N.J.G.J., A. Jayasinghe, and W.A.S.S. Dissanayake, *Progress Report: Current Status of Plastic Packaging Materials in Sri Lanka*, 2010, University of Sri Jayewardenepura: Nugegoda, Sri Lanka. 123, 124

[36] Petley, D., Garbage dump landslides, in *The Landslide blog*, 2008, American Geophysical Union.

[37] Scheirs, J., Additives for the Modification of Poly(Ethylene Terephthalate) to Produce Engineering-Grade Polymers, in *Modern Polyesters: Chemistry and Technology of Polyesters and Copolyesters*, J. Scheirs and T.E. Long, Editors. 2003, John Wiley and Sons: Chichester, England. 124

[38] Van Zon, L. and N. Siriwardena, *Garbage in Sri Lanka - An Overview of Solid Waste Management in the Ja-Ela Area*, 2000, Integrated Resources Management Programme in Wetlands (IRMP): Colombo. 124

[39] Steele, P., A. Gunawardena, and B. De Silva, *Study on Post-Consumer Plastics in Sri Lanka*, 2005, Environment and Management Lanka Ltd. 124

[40] Ministry of Environment, *Database of Solid Waste*, 2005, Sri Lanka: Ministry of Environment, Government of Sri Lanka. 125, 128

[41] Asian Institute of Technology (AIT), Solid Waste Management in Asian Countries: Case Studies, in *Sustainable Solid Waste Landfill Management in Asia*, Asian Institute of Technology: Pathumthani, Thailand. 125

[42] Kent, M., Development of a Social Impact Assessment methodology and its application to Waste for Life in Buenos Aires, in *School of Environmental Systems Engineering*, 2010, The University of Western Australia: Perth. pp. 88. 125

[43] Kamburugamuwa, A., Sri Lanka urged to better connect rural areas to Colombo, in *Lanka Business Report*, 2010, Colombo.

[44] Perera, K.K.C.K. et al., Assessment of sustainable energy potential of non-plantation biomass resources in Sri Lanka. *Biomass and Bioenergy*, 2005. **29**: pp. 199–213.
DOI: 10.1016/j.biombioe.2005.03.008 126, 127

[45] Chan, E. and C.R. Elevitch, *Cocos nucifera (coconut)*. Species Profiles for Pacific Island Agroforestry, 2006. 126

[46] van Dam, J.E.G., Coir Processing Technologies - Improvement of Drying, Softening, Bleaching and Dyeing Coir Fibre/Yarn and Printing Coir Floor Coverings, in Technical Paper No. 6, 2002, *Food and Agriculture Organization of the United Nations and Common Fund for Commodities*, Amsterdam. 127

[47] Industrial Development Board of Sri Lanka, *Developing the Coir Sector in North-Western Province: Value Chain Development for more Competitiveness and Decent Work*, 2007, Industrial Development Board of Sri Lanka and Industrial Services Bureau. 127

[48] Lantin, R. *RICE: Post-harvest Operations*. Post Harvest Compendium 1999 14 October 1999; Available from: http://www.fao.org/inpho/inpho-post-harvest-compendium/cereals-grains/en/. 127

[49] Yang, H.S., et al., Rice husk flour filled polypropylene composites; mechanical and morphological study, *Composite Structures*, 2004. **63**: pp. 305–312.
DOI: 10.1016/S0263-8223(03)00179-X 127

[50] Yang, H.S., et al., Effect of compatibilizing agent on rice husk flour reinforced polypropylene composites, *Composite Structuctures*, 2007. **77**(45–55).
DOI: 10.1016/j.compstruct.2005.06.005 127

[51] Nourbakhsh, A., F.F. Baghlani, and A. Ashori, Nano-SiO$_2$ filled rice husk/polypropylene composites: Physico-mechanical properties, *Industrial Crops and Products*, 2011. **33**(1): pp. 183–187. DOI: 10.1016/j.indcrop.2010.10.010 127

[52] Yang, H.-S., et al., Properties of lignocellulosic material filled polypropylene bio-composites made with different manufacturing processes, *Polymer Testing*, 2006. **25**(5): pp. 668–676.
DOI: 10.1016/j.polymertesting.2006.03.013 127

[53] Ndazi, B.S., et al., Chemical and physical modifications of rice husks for use as composite panels, *Applied Science and Manufacturing*, 2007. **38**(3): pp. 925–935.
DOI: 10.1016/j.compositesa.2006.07.004 127

[54] Senanayake, D.P., U. Daranagama, and M.D. Fernando, Availability of paddy husk as a source of energy in Sri Lanka, in *Research Seminar*, 1999: NERD Center, Sri Lanka. 127

[55] Ricegrowers" Association of Australia. *The Rice Growing and Production Process*,; from: `http://www.aboutrice.com/downloads/rice_growing.pdf`. 127

[56] Liyanapathirana, R., Comparative Analysis of Rice Marketing System in Sri Lanka - Pre and Post Liberalization Period, in *Departmet of Agricultural Marketing Co-operation and Agribusiness Management*, 2006, University of Agricultural Sciences: Dharwad. 127

[57] Pan, M.-Z., et al., Effects of Rice Straw Fiber Morphology and Content on the Mechanical and Thermal Properties of Rice Straw Fiber-High Density Polyethylene Composites, *Journal of Applied Polymer Science*, 2011. **121**: pp. 2900–2907. DOI: 10.1002/app.33913 127

[58] National Cleaner Production Center, Report II: Waste Biomass Quantification and Characterization, in *Project on Converting Waste Agricultural Biomass to an Energy/Material Resource*, 2010, United Nations Environment Programme: Sri Lanka. 127

[59] Perera, A., *Pelwatte Sugar Industries Limited*, 2010, Asha Phillip Securities Ltd: Colombo. 128

[60] W.R.G. Witharama (pers. comm), Sugar industry in Sri Lanka, July 19, 2011. 128

[61] Monteiro, S.N., et al., Sugar Cane Bagasse Waste as Reinforcement in Low Cost Composites, *Advanced Performance Materials*, 1998. **5**: pp. 183–191. DOI: 10.1023/A:1008678314233 128

[62] Acharya, S.K., et al., Weathering Behavior of Bagasse Fiber Reinforced Polymer Composite, *Journal of Reinforced Plastics and Composites*, 2008. **27**(16–17): pp. 1839-1846. DOI: 10.1177/0731684407082544 128

[63] Koopmans, A. and J. Koppejan, Agricultural and Forest Residues - Generation, Utilization and Availability, in *Regional Consultation on Modern Applications of Biomass Energy*, 1998, Food and Agriculture Organisation of the United Nations: Kuala Lumpur. 128

[64] Department of Census and Statistics, *Municipal Solid Waste Statistics 1998*, 2001, Colombo: Department of Census and Statistics, Government of Sri Lanka. 128

[65] Bandara, N.J.G.J., Municipal Solid Waste Management—The Sri Lankan Case, in *Conference on Developments in Forestry and Environmental Management in Sri Lanka*, 2008, Kalatura, Sri Lanka. 128

[66] Asian Development Bank, *Country Partnership Strategy: Sri Lanka 2009–2011*, 2008, Asian Development Bank. 128

[67] Board of Investment, Free Trade Zones and Industrial Parks, available from: `http://www.boi.lk/free_trade_zones_industrial_parks.asp`. 128

[68] Sivananthiran, A. Promoting decent work in export processing zones (EPZs) in Sri Lanka. 128

[69] Aparakkakankanamage, A., Globalization, Sustainable Development, and Environmental Problems in the Third World: A Case Study of Sri Lanka, in *Department of Government and Politics*, 2005, University of Maryland. 128

[70] Geocycle, *Perception Survey*, 2009, University of Sri Jayewardenepura, Holcim (Lanka) Ltd.: Nugegoda, Sri Lanka. 128

[71] Georgia Gulf Chemicals and Vinyls, LLC. Material Safety Data Sheet: PVC Compounds, *Material Safety Data Sheets*, 2005; available from: `http://www.qubicaamf.com/QubicaAMF/files/1f/1f7a12e4--2e3a-4079-bbcc-3b21ed005178.pdf`. 129

[72] Denka Singapore, Pte Ltd. Material Safety Data Sheet: Polystyrene, 2003; available from: `http://www.southlandpolymers.com/pdf/polysty/MSDS%20MF21--301%20%282%29.pdf`. 129

[73] Environment Australia, Carbon monoxide, in *Air toxics and indoor air quality in Australia: State of knowledge report*, 2001, Department of Sustainability, Environment, Population, Water and Communities, Commonwealth of Australia: Canberra. 129

[74] Beer, T., et al., Atmosphere: Theme commentary, in *State of the Environment*, 2006, Department of Sustainability, Environment, Population, Water and Communities, Commonwealth of Australia: Canberra. 129

[75] Department of Sustainability, Environment, Water, Population and Communities (DSEPWC). Hydrochloric acid: Environmental effects, *National Pollutant Inventory*; available from: `http://www.npi.gov.au/substances/hydrochloric-acid/environmental.html`. 129

[76] Pervaiz, M. and M.M. Sain, Carbon storage potential in natural fiber composites, *Resources, Conservation and Recycling*, 2003. **39**: pp. 325–340. DOI: 10.1016/S0921-3449(02)00173-8 130

Authors' Biographies

RANDIKA JAYASINGHE

Randika Jayasinghe is a second-year Ph.D. student at the School of Environmental Systems Engineering, University of Western Australia. She is an AusAID Leadership Awards Scholar, working towards developing poverty reducing solutions for sustainable waste management in Sri Lanka. Randika has carried out many projects in the waste management sector in Sri Lanka and is interested in studying how social and environmental justice principles can be integrated into waste management in developing countries.

USMAN MUSHTAQ

Usman Mushtaq is interested in issues at the intersection of justice, technology, and engineering. He recently finished with his academic work and is delving into the professional world. He is particularly interested in looking at how the public is consulted and engaged in decision-making around engineering projects and environmental impact assessments.

TONI ALYCE SMYTHE

Toni Smythe studied environmental engineering at the University of Western Australia and became involved in Waste for Life through her final year thesis, which assessed the feasibility of a new project in Sri Lanka. She is currently working for the Department of Water, Perth, within the surface water hydrology team.

CAROLINE BAILLIE

Caroline Baillie is Chair of Engineering Education for the Faculty of Engineering, Computing and Mathematics at the University of Western Australia. Caroline is particularly interested in ways in which science and engineering can help to create solutions for the environment as well as social problems. She is Editor of the Morgan and Claypool series, 'Engineers, Technology and Society," and Director of the Australian program (funded by the Office of Learning and Teaching) "Engineering Education for Social and Environmental Justice" of which this book is a part. She founded the global "Engineering and Social Justice" network (www.esjp.org) and applies this lens to her work for "Waste for Life" (http://wasteforlife.org/) which she co-founded with Eric Feinblatt.

Printed in the United States
by Baker & Taylor Publisher Services